Technology, Social Change and Human Behavior

"Using technology to crush the cycle of greed and inertia and help us achieve a state of 'Augmented Humanity' founded in generosity, compassion, honesty and courage is a beautifully hopeful idea. In these anxious and uncertain times, it offers a much needed, enticing and ultimately achievable vision of a new utopia. Reading this book makes you feel better, but it does much more. It takes you on a journey of imagination, from the closed-mindedness of hyper-individualism to the joys of authentic human connection, with self and others. Read it and you will discover more of who you truly are. In doing so you will feel more awake, more purposeful and more alive."

—Gareth Owen OBE, *Humanitarian Director, Save the Children UK International, London*

***

"Cornelia Walther's book arrives at an important time as Covid19 has accelerated the digital transformation of many elements of our society. Using technology to encourage humans to embrace digital technology and use it for social change is a brilliant concept and one that needs to be explored further. Never has the world been more connected and yet strangely at isolated at the same time. The author explores how technology could be used to bring about real social change at a time when a global population is looking for direction."

—John E. Roche, *CMC. Global Strategy Director, Enablement. SAP Premium Services. New York*

***

"The truth is that we live in a networked world, interconnected. Thus, the challenge is to ensure that technological advance is at the service of improving people's quality of life. This is possible if we consider that personal well-being is the result of collective well-being and then no one can achieve their well-being in isolation from what happens in their context. This book invites you to discover how to enhance social change dynamics, and in this sense improve people's quality of life, using technology overtime to make it sustainable. If you are interested in these topics, you cannot stop reading this new and compelling book by Cornelia Walther."

—Dr Graciela H. Tonon, *Director Master of Social Sciences and Center for Research in Social Sciences (CICS-UP). Faculty of Social Sciences. University of Palermo, Buenos Aires*

***

"*Cornelia Walther takes a refreshingly original look at the relationship between technology and social change, incorporating the Poze methodology, which she developed after working for UNICEF and the World Food Program in a number of countries, including Haiti, Chad, Afghanistan and the Democratic Republic of the Congo. Her latest book is an invaluable tool for promoting tolerance and societal transformation throughout the world.*"

—Dr. Joann Halpern, *Director, Hasso Plattner Institute, New York. Adjunct Professor of International Education, New York University*

***

"*Imagine a world where technology is designed to elevate your consciousness, point you towards your highest self and most noble aspirations. This is not about the future; this technology is already here, but what intention do we collectively as a species bring to our creations? Our challenge today is to understand that with every choice we make to adapt to new technology, our world is changed. Thus, it becomes increasingly important to make sure that our choices are mindful. This book opens a world of possibilities and poses important questions about who we are and how we can be nudged to interact with technology as it evolves. It is essential reading for the curious mind and highly recommended for those among us with the power to shape the future of the technology that will shape how we evolve as a species.*"

—Marielle Sander, *UNFPA Representative. Papua New Guinea. Port Moresby*

***

"*The future is uncertain. Nevertheless, with solid theory and a good model, it is possible to peek and discern probable trends. This book provides a superb model that not only does exactly that but also delivers tools and practical ideas to use technology for the betterment of life.*"

—Enrique Delamonica, *Senior Adviser Statistics and Monitoring (Child Poverty and Gender Equality), United Nations Children's Emergency Fund (UNICEF)*

***

Cornelia C. Walther

Technology, Social Change and Human Behavior

Influence for Impact

Cornelia C. Walther
POZE
Tuebingen, Germany

ISBN 978-3-030-70001-0   ISBN 978-3-030-70002-7   (eBook)
https://doi.org/10.1007/978-3-030-70002-7

This Palgrave Macmillan imprint is published by the registered company Springer Nature Switzerland AG
The registered company address is: Gewerbestrasse 11, 6330 Cham, Switzerland

PREFACE

The central argument of this book is that society can be shaped to lift individuals to fulfill their potential, if those who constitute Society (Us) aspire to that objective and act accordingly. Technology can help us to make that happen. It cannot do it for us. To serve the Common Good[1] technology must be designed, delivered, and used with this goal in (the human) mind. That feedback loop of human aspirations and artificial intelligence is one of the central messages of this book. It is not a technical book and the cited examples are not an extensive inventory of the technological prowess that the world disposes of. The aim is not to draw an inventory but to show certain patterns that affect society.

Humanity is ready to take a leap from bigger to better to brighter; from catering to a culture of cash and casual carelessness to collective consciousness. This leap is not quantitative but qualitative. Societies that nurture quality of life for all while protecting the environment become possible when humans adopt the understanding that their own wellbeing, and that of others, is not only connected but represents two sides of the same coin. A state of Augmented Humanity (AH) is within reach when individuals realize that sharing and caring enhance their own lives, and act accordingly because they want to.

Technology can contribute to the concretization of AH in our lifetime. But, to nurture AH, technology must be both designed and used with a particular mindset. The outcomes of technology depend on the attitudes and aspirations that underpin it. Fire can serve to cook a meal, or to burn a village—the end is not conditioned by the mean itself. Furthermore, it is merely a mean to an end not an end by itself. Impressive as the technological progress is, it does not free us from the responsibility of making the right choices.

We shape the technology that shapes us. Millions of customers and CEOs, consumer consortia and communities have been pondering which impact Artificial Intelligence (AI) and ancillary technologies will have on our future; whether they will be beneficial or detrimental. The answer to that interrogation does not reside with the sophistication of algorithms, nor the development of ever more capable devices or the mining of ever bigger quantities of data. The impact of technology on this (and future) generations and on the Planet depends on those who live, think, and feel—and act, today (Us).

To optimize the positive influence of technology on individuals and institutions, the intentions that precede the design and implementation of hard and software must be geared toward collective progression, not personal profitability. Conceived with the understanding of interconnected impact, technology may compensate for limitations of human nature; starting with our tendency to biased judgments, willful blindness, inertia, and greed. As we will see over the course of this book, technology can accompany the human journey to a higher self. It represents neither the path nor the traveler.

Depending on the mindset of engineers and entrepreneurs, of users and utility managers technology may be refined ever further. We have just begun to explore the potential of mind-oriented technology; the question is to move from Artificial Intelligence (AI) to Artificial Emotional Intelligence (AEI) to Aspirational Algorithms (AA) in order to nurture a state of Augmented Humanity (AH). The last stage entails machines which are programmed with human values and directed toward the achievement of human aspirations. Undeterred by greed and jealousy, AA endowed machines can influence, or take over, decision-making processes in which humans traditionally succumb to their lower instincts. In other words, technology can help us on the path to AH if it is designed with that purpose—however, it will not be designed with that purpose unless we are already on that path. This is not a tautological argument, nor a catch-22.

Technology Can Crush the Cycle of Greed and Inertia—If We Design It to Do So

Well-intentioned individuals and institutions have tried for centuries to stop and reverse trends that lead to an imbalance of needs and means—poverty. The status quo, in which, globally, inequity has been on the rise for decades, illustrates that their efforts have at best alleviated the suffering that derives from that situation. More candidly spoken—the thinkers and doers of the past and present failed to end avoidable deprivation. Blaming 'the system' for this failure means falling prey to the chicken and egg conundrum. Cause and consequence often appear indistinguishable but blaming 'the system' equals shaming us, individually and collectively, as we perpetrate the 'system' that we are part of. Aspirational Algorithms (AA), a type of technology designed to nurture the common good, *may* help us to break the cycle and move toward AH.

A futuristic scenario where the world is ruled by robots is often portrayed as an apocalypse. To be fair in this debate, it is worth looking at the world that is emerging from millennia of a human regime. The pace of human evolution has been extraordinary. We know and possess more, while living longer to enjoy it. However, war and pollution, climate change and the exploitation of people and planet meander in parallel to that progress.[2] Nature is declining at rates unprecedented in human history, with up to 1 million species at risk of extinction due to human activity (Kailash and Lambertini 2020). The World Economic Forum's Global Risks Report 2020 ranked biodiversity loss as one of the top five risks in terms of impact and likelihood over the coming decade. For the first time in history, humans are the drivers of climate and environmental change—we live in the Anthropocene, with devastating consequences. If the current rate of destruction continues unabated, some biomes (e.g. tundra, grasslands, coral reefs, forests) will cross irreversible tipping points. The impact may make the world uninhabitable for humans. This book does neither argue in favor of robotic empires nor advocate to outlaw machines. It merely looks at options to make the best of a future where technology will be part of our life.

Reversing the current trend starts with a shift in human perspective—from closed-mindedness to connection, from constraining mindsets to organic changes, from categories that parcel and judge experiences to a continuum of humanness. A shift from competition to complementarity. Much good has happened over the past centuries and is happening right now. Much more is needed, quickly. We have a solid point of departure, both in terms of means and motivation. We must not wait while the momentum slips away.

The technological assets that we now take for granted would have appeared like magic centuries (even decades) ago—from cellphones over computers down to AI. The present pages seek to disperse the magical dust to distill the human outcomes. Paradoxically, technology can serve us in these endeavors if it is built and used by humans who aspire to that outcome. If human engineers and users do justice to their innate ability of generosity, compassion, honesty, and courage, technology will reflect these traits.

Individual wellbeing is the cause and consequence of collective welfare. One without the other is unsustainable. Together they condition the transition from now to Augmented Humanity.

Anchored in the understanding that everything operates in a constant interplay which can be influenced, we look at intrapersonal dynamics, from small to large-scale flows. Change starts from the inside out with the aspiration of human beings. It is nurtured from the outside in, with action that makes the best use of tools and techniques. The POZE paradigm[3] offers a multidimensional perspective that can be used to understand and promote social change, via personal transformation and vice versa.

The changeover involves 4 stages that complement each other: A new PERSPECTIVE on who we are, individually and collectively; which enables us to pursue an OPTIMIZATION of the factors that condition our behavior, as well as the influence of these factors on technology and conversely the influence of technology on them. Once a fresh dynamic is initiated it gradually dismantles the cycle of inequity. The journeys of personal change and social transformation propel each other; taking the individual traveler and humanity altogether forward to the best version of itself, a ZENITH. We may never reach the final stage; because once we reach what we thought was the ultimate goal, we find that the finishing line has shifted. Rather than one zenith, there are many; we grow to grow further, from one milestone to the next. Along the way we come closer to our highest self. As we learn more about ourselves our perspective on life changes; mental filters and misperceptions dissolve. The consequence is EXPOSURE to reality as it is, including our own nature. This book looks at the 4 phases that are involved in this process:

Chapter 1 establishes the parameters of a PERSPECTIVE that is used in the rest of the book. It presents the 4 dimensions that underpin individual existence (soul, heart, mind, body), the 4 dimensions that determine society (individuals, communities, countries, planet), as well as 4 principles that rule throughout these dimensions (connection, change, continuum, complementarity). *Awareness* of their existence and their mutual influence conditions the outcomes of this journey. Once we know what influences us, we can shape that influence in view of the desired outcomes.

Chapter 2 investigates technological advances that may be conducive or detrimental to the OPTIMIZATION of individual and collective life-determinants. We look at the role that recent advances in high-end technology may play to accelerate positive influence within and among multiple dimensions. Because *Acceptance* of the dynamics that underpin such influence conditions their optimization, which determines the gradual fulfillment of our potential.

Chapter 3 goes into the undercurrents that influence the move from complacency to compassion to change, or the ZENITH of humanity. Certain aspects of our unexplored humanpotential in the context of high-tech changes will be looked at from the micro (individual), meso (community), macro (country) and meta (global) angles. *Alignment* is investigated as both, a process and an outcome. Because the gradual alignment (process) of the dimensions that shape our individual experience eventually leads to the *Alignment* (outcome) of individuals among each other, and thus the overall synchronization of society. It should be mentioned that the word 'ZENITH' is used here not as an ultimate point, but one of many successive peaks along a path of continued improvement. Over the course of our individual existence, as well as from one generation to the next, many ZENITHs can be pursued—as with each stride "the head above the line shifts, and the horizon expands" (Corominas 1987).

Chapter 4 concludes with a set of recommendations to increase the likelihood of positive outcomes. The latter depend on EXPOSURE, to reality as it is; including our own human nature. Knowledge is power, and the latter entails *Accountability*. Once we are aware of the circumstances that influence our life, we are responsible to use that awareness to optimize our internal and external influence for the betterment of society. Awareness of the dynamics that underpin who we are, combined with the aspiration of the common good, makes it then possible to create Aspirational Algorithms (AA); a type of technology that maximizes the benefits for all parties. The concluding recommendations relate to the feedback-loop that causes and conditions human aspirations and aspirational technology.

Differently said, the first chapter offers a new perspective or framework to understand human beings and society. The second chapter looks at technology, pointing out certain cornerstones of the tools that are of interest in the ambition of optimizing social outcomes. The third chapter brings both together and illustrates how the interplay of technology and humanity may be beneficial. The fourth chapter concludes with a set of suggestions that put human beings and Planet Earth front and center in a future with technology. While the accent is herein put on offline/low-tech solutions, two illustrations of the type of technology that makes the best of artificial and natural assets for the common good of humanity are offered.

The symmetry that underpins the POZE paradigm is reflected in the structure of this book. It connects the different stages of personal change (awareness, acceptance, alignment, accountability) to the 4 pillars of social transformation (perspective, optimization, zenith, exposure). These two sets of elements mirror the 4 dimensions of every individual (soul, heart, mind, body) which relate to 4 complementary values that matter for sustainable change (generosity, compassion, honesty, courage). The schema below captures the terms that will be used in this book and their interconnected logic. It is a snapshot that will be unpacked as we move through the chapters (Fig. 1).

Tuebingen, Germany   Cornelia C. Walther

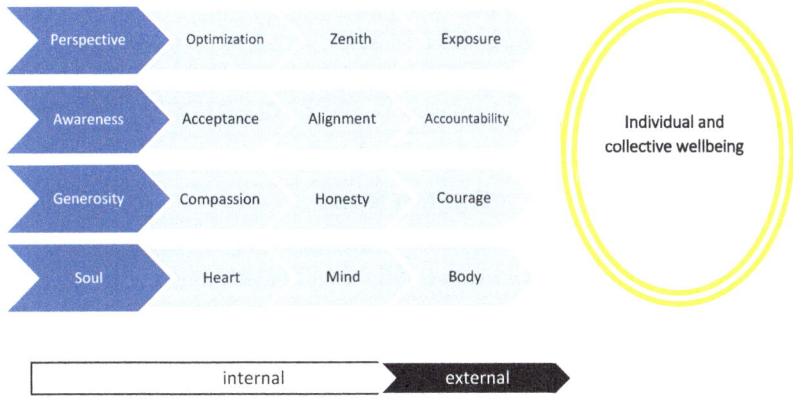

Fig. 1 Interrelated stages of being and becoming. The symmetry that under-pins the POZE paradigm is reflected in the logic that connects different stages of personal change (awareness, acceptance, alignment, accountability) to 4 pillars of social transformation (perspective, optimization, zenith, exposure). They mirror the 4 dimensions of every individual (soul, heart, mind, body). In addition, they echo 4 complementary values that matter for sustainable change (generosity, compassion, honesty, courage)

NOTES

1. In the present context we refer to the Common Good as that as attitudes and actions that benefit society as a whole, in contrast to the private good of individuals and sections of society (Encyclopedia Britannica).
2. This is in line with the paradox of progress that as stated by Nell (2019) and described almost 150 years ago by George (1879).
3. POZE is based on the understanding that change starts from the inside out and is nurtured from the outside in—as we will see in Chapter 1. POZE has different meanings, it: (i) represents the 4 outcomes of the logic that under-pins it (Perspective, Optimization, Zenith, Exposure); (ii) is an acronym that encompasses the 4 core concepts of the paradigm (Purpose, Om, Zoom, Expression); (iii) translates as 'inner peace' from Haitian Creole, country where the dynamic began in 2017; and (iv) illustrates an exer-cise to nurture inner peace daily (Pause, Observe, Zoom in, Experience). Walther 2020-a. Furthermore, it stands for is the description of a move-ment: People On Zenith Exploration. And, as will be seen in Chapter 4, it reflects the layers of optimization covering Purpose orientation that tran-scends personal interest, Optimism for the future; Zeal to achieve the best

outcomes for humanity; and Exploration of opportunities with the ambition to move beyond forgone conclusions. These relate to the overall aim of OPTIMIZATION, yet they go deeper by establishing a more granular mental matrix.

References

Corominas, J. (1987). *Breve diccionario etimológico de la lengua castellana* (in Spanish) (3rd ed., p. 144). Madrid. ISBN 978-8-42492-364-8.

Encyclopedia Britannica. *The common good*. Retrieved November 2020, from https://www.britannica.com/topic/common-good.

George, H. (1879). *Progress and poverty: An inquiry into the cause of industrial depressions and of increase of want with increase of wealth*. New York: Doubleday & McClure Co.

Kailash, D., & Lambertini, M. (2020). *Why 2020 is the year to reset humanity's relationship with nature*. Retrieved November 2020, from https://rb.gy/iqfs3m.

Nell, E. (2019). *Henry George and how growth in real estate contributes to inequality and financial instability*. Palgrave Macmillan.

Acknowledgements

This book has organically arisen from the past four decades of my life, and the millennia that preceded them. It combines past and future, applying the POZE paradigm to an issue that is all-pervasive, and influences us mostly under the radar. Technology is everywhere. It has been illustrating human ingenuity since the onset of times, with a slim line to separate danger from delight, and destruction from realizing the dream of a better Society.

Every generation felt like the cusp of innovation, the peak of humanity—and always more was to come. Today technology confronts us yet again with the onset of radical transformation. And with it, the challenge to acknowledge and reconcile the two streams of human nature—personal interest and the common good. Whether we zoom in on the former or zoom out to embrace the latter, our choices influence the future of technology and thus the course of society from hereafter.

Our understanding of the world is an organically evolving kaleidoscope, ever changing. Thus, I am writing this book to honor many individuals—those who travelled the road of questions before me; who walked with me; and who will come later.

I am indebted to the thinkers whose discoveries led to the technology that this book is about, with gratitude to those who remained unacknowledged because they worked backstage. Scientific exploration would not be possible without the unnamed who produce the hardware, the coders who translate overarching objectives into everyday scripts, the research assistants who test and refine. Their efforts and sacrifices happen in silence, yet the sum of their contributions is an inherent part of the scientific prowess that we admire today.

It is impossible to list all of those who helped me along my journey in writing this book. However, my parents, Barbara and Manfred Walther, must be named as I would not be who I am without them being who they are. Furthermore, I am grateful to Enrique Delamonica for his kind and candid review of the manuscripts that led up to this book. Lastly, I am grateful to Alina Yurova and Anne-Kathrine Bircher my fabulous editors at Palgrave Macmillan and their entire team, especially Aishwarya Balachandar for their continued support.

Finally, and foremost, Thank You—for choosing this book. Time is rare, and mental real estate precious. I appreciate that you give me a chance. Without you the content of these present pages will be forgotten soon, drifting away with the trickle of time. If you find the present propositions valid, your action can make a difference for the dissemination of the present perspective, and for those whose life is on the line as technology moves ahead in gigantic strides. Us.

As always, I am eager to hear your feedback. Please get in touch via LinkedIn or via the Contact section on https://www.poze.cc.

SCOPE

This book looks at the changing continuum that links individuals, communities and society. An outline of Aspirational Algorithms (AA) and Valuable Wearables is presented as tools to shift from an AI culture to the cultivation of Augmented Humanity (AH).

The human mindset that is behind the design and use of technology determines the outcomes of technology. If the intended outcome is the common good, then the preceding human aspiration must be geared towards that goal. Only a technology that is conceived with the aspiration of a society that lifts individuals to fulfill their potential can be a gamechanger for good.

Based on an understanding of the constant interplay between the 4 levels of human existence—soul, heart, mind, body, expressed as aspirations, emotions and thoughts and sensations—technology may serve to systematically sway individuals from inspiration to desire, from informing to the ignition of tangible transformation. This transition is explained along the Scale of influence.

Two convergent and mutually influencing dynamics are analyzed: (1) the influence of values and aspirations on the impact of technology; and (2) the influence of technology on the attitude and action of users, and thus the Society they evolve in. Both assess how hardware and software can serve everyone to live a meaningful happy life.

CONTENTS

About the Author

Cornelia C. Walther combines praxis and research. As a humanitarian practitioner, Cornelia worked for nearly two decades with UNICEF and the World Food Program in large-scale emergencies in West Africa, South Asia and Latin America, mostly operating as head of communication. As coach and researcher, she collaborates since 2018 with the Center for humanitarian leadership at Deakin University and serves as a mentor within the Harvard Women in Defense and Diplomacy network. Being part of the European Union's Network on humanitarian assistance (NOHA), she lectured for five years at Aix-Marseille's Law faculty.

Aside from her interest in the multiple shapes of influence, Cornelia's focus is on social transformation from the inside out, looking at individual aspirations as the point of departure. In 2017, she initiated the POZE (*Purpose, Om, Zoom, Expression*) dynamic in Haiti, offering individuals tools to identify and pursue their aspirations. The network is now expanding into the Americas, Africa, and Europe. Her objective is to refine a methodology that influences people toward *wanting* to get involved in social change processes, rather than obliging them to act for the sake of others. She holds a Ph.D. in Law and is a certified yoga and meditation teacher.

Books published by Macmillan Palgrave/Springer in 2020, include '*Development, Humanitarian Action and Social Welfare*'; '*Humanitarian*

Work, Social Change and Human Behavior'; and '*Development and Connection in Times of COVID*'. For further details see https://www. poze.cc.

Get in touch with her via https://www.linkedin.com/in/corneliaw alther.

LIST OF FIGURES

CHAPTER 1

Perspective

Abstract Our physical state influences our thoughts and emotions, whereas the latter influence our physical health and overall wellbeing. The POZE paradigm which underpins this book is based on the understanding that human existence is a composition of soul, heart, mind, and body expressed as aspirations, emotions, thoughts, and sensations. Their interaction stands in symmetry to the interaction of the 4 dimensions of society—individuals, communities, countries, and Planet Earth (Walther 2020a). This multidimensional logic and the 4 universal principles that apply within and throughout both will be looked at here. Awareness of these multi-dimensional dynamics is the first of 4 stages required for personal change and hereby for sustainable, large-scale transformation.

Keywords Perspective · Paradigm · Connection · Continuum · Complementarity · Awareness · Influence

The interplay between everything and everyone can be influenced and optimized. Technology can support that process—not only at the level of the individual, but systemically, thus large scale. Technology may serve to realize a society that is anchored in the shared ambition of lifting individuals to fulfill their potential. For this to happen, it must be designed and delivered with this ultimate outcome in mind

1

C. C. Walther, *Technology, Social Change and Human Behavior*, https://doi.org/10.1007/978-3-030-70002-7_1

We start here with the principles that rule our existence, followed by the multiple dimensions that underpin who we are—as a person and as a society. These parameters underpin the PERSPECTIVE that is developed in this book. *Awareness* of them conditions the systematic optimization of human experiences and expressions. We can intentionally influence only what we know.

Everything is connected. In the same way in which our devices shape the speed and power of the software that is used on them, our physiological set-up influences how and what we think and feel, and hereby what we do. Conversely, similar to software upgrades that impact the design of new hardware due to changed requirements and possibilities, our inner setting influences our experiences and expressions in the material, external surroundings. Simultaneously, both evolutions influence our expectations, which impact our perception of 'reality' and thus our reaction to it.

Individually and collectively, we are setting the path that we are walking. Each step influences the next. Whether we like the destination that we end up at depends, thus, less on our point of departure, than on our choices along the way. It is never too late to change directions. To make the right decisions and turn at the most promising crossroads, it is useful to have a reliable weather forecast and a solid grasp of the landscape that we are navigating; to understand the equipment that we have taken along on the journey; and to comprehend our co-travelers. A journey rarely goes exactly as planned even when those factors are well known; yet even more rarely, it reaches the desired goal if the traveler does not know what that goal is.[1]

Tomorrow, the next month, and the next decade are unknown. This opaque setting of life has been a constant forever, playing a fundamental role in everything, from personal life-planning to the understanding of economic cycles and stagnation.

Though the principle of uncertainty is well known by historians and economists alike, scientists, investors, and each of us persist in trying to predict future outcomes based on past experiences. An endeavor that is doomed to fail. "The inability to predict outliers implies the inability to predict the course of history" (Taleb 2007). COVID-19 illustrated, yet again, how bad we are at predicting the future, and worse, how badly equipped we are to prepare for future outcomes that are likely to arise, yet are undesirable. Wishful thinking and chosen blindness are two mental features that we will look at closer in the coming pages.

Amid the all-pervasive uncertainty that surrounds us are constant parameters that can be understood and used as tectonic plates. These constants are the 4 components of our being, soul, heart, mind, and body—which echo the collective arena, composed of micro-, meso-, macro-, and meta-dimensions, in the form of individuals, communities, countries, and planet earth. Paradoxically, we usually ignore the interplay of these parameters even though we grow up with them as our constant companions. Building a safe island amid the flow of change, POZE offers a 4-dimensional perspective which connects the known parameters within a coherent framework. The visible and the invisible, the inside and the outside, mind and matter, individual and society all function along the same principles. The PERSPECTIVE of these principles and dimensions offers a minimum of stability. Because once awareness of the building blocks is established, one can start to systematically optimize the interplay between these blocs, undeterred by factors that are out of our immediate control.

1.1   PRINCIPLES

Individual wellbeing is the cause and consequence of collective welfare, due to 4 principles that influence both spheres: change, connection, continuity, and complementarity.

Life is the result of a logic that precedes human laws and regulations. Pointing it out here is not done with the aim of changing it. Efforts invested in this endeavor are futile. We do not learn about gravity at school to reverse it. Knowing why an apple falls to the ground when dropped does not lead us to smash apples onto the carpet to make them fly. Knowing *why* they fall does not prevent them from falling, but it helps us understand why it is useful to be cautious when carrying a porcelain vase to the sink. Getting an idea of the 4 principles that underpin life is useful in a similar perspective. We can cope best with circumstances that we comprehend. Once we know what we deal with we can adjust. Unless we chose to ignore it, or to consciously move against the prevailing logic (in a self-destructive manner). The following 4 principles influence all that is:

Connection—Everything is linked to something else. Nothing happens in a vacuum. The multiple interplays that we are part of render our existence a domino-chain. Whatever we do, small and large has consequences.

It has ripple effects that influence our being and becoming, directly and indirectly.

Change—Everything is always evolving. Nothing remains as it is. Evolution is part of who we are and what surrounds us. Accepting this dynamic of change as a given is the first step toward benefits from this organic set-up of life. Holding on to property, positions, and people; trying to ascertain circumstances; attempting to control what is inherently fluid (mis)directs energy that can be put to better use.

Continuum—Everything is part of a whole. Nothing is completely discrete. Separation is an illusion. Time, space, and situations occur along an ongoing trajectory that connects mind, matter, moments, and movements within a spiral dynamic.

Complementarity—Everything needs something else to be complete, while representing the missing part of something other. Nothing is complete in isolation. Every individual, every animal, and every part of nature contains a unique set of assets that enable it to help other(s) to be fulfilled. We are part of a symphony with countless notes that are intrinsically attuned to one another. The sum is more than its parts because of the interplays that join these parts within a totality. The magic of the whole unfolds when all parts are synchronized.

1.2   INDIVIDUALS

The foundation of life is a twice 4-dimensional dynamic. Individual existence is a composition of 4 dimensions (soul,[2] heart, mind, and body). Equally the collective arena is composed of 4 dimensions—micro, meso, macro, and meta (individual, community, country, planet).

The experiences and expressions of every person are conditioned by 4 dimensions which constantly interact and influence each other, and hereby influence the interaction of the individual with the outside. What happens in one dimension has repercussions in the others.

- The **Soul** is the core of our being. Identifying and fulfilling the purpose of our existence is the beginning and closing of life. In the process of living, we come closer to our highest self, which allows us to share that refined version of ourselves with others. Once we have thus come into the proximity of our best self, the desire to care and share arises naturally. Daring to start that journey is the most challenging part of the way.

- The **Heart** keeps us alive, metaphorically and in real terms. Every beat of the heart is a minuscule but fundamental part of this intrinsically complicated process that maintains our organism. Similarly, our emotions are the warm and relatable part of our being that makes us a social being who loves and loathes. Emotions run underneath our intellectual deliberations and decision-making processes. They play a major role in our choices.
- The **Mind** filters, analyzes, and judges the sensorial inputs that enter via the physical structure. Based on past experiences, it draws conclusions and makes decisions. While emotions arise and pass beyond our control, thoughts are subject to conscious amendment. Thoughts result in emotions; these thoughts offer an entry-point to get a grip on the resulting emotions. Gaining control over the flow of thoughts begins with awareness of their existence and influence.
- The **Body** is the physical interface between mind and matter, connecting our internal and external realm. It is the channel through which individuals experience their environment and express themselves in it. The body is at the same time conductor and communicator when it comes to the interaction with the environment. It is the (double) filter that influences how we perceive other people, and how we are perceived by them. It carries out the decisions that were processed in the mind, reflecting the underpinning aspirations and emotions.

Every dimension influences the others and is influenced in return. Experience is subjective, deriving from the inner set-up of the person who goes through the situation. Our thoughts and emotions at the time of the situation affect our perception of that situation. Deriving from our past via memories, belief systems, and background, they impact how we analyze the present sensorial intake (Kahneman 2011).

How we experience our environment influences how we express ourselves in it. Conversely how we express ourselves in a situation influences our experience. It is an ongoing looping from the inside out and from the outside in; from the past to the present; and from here into the future. Every action, even those taken without firm conviction, leaves traces in the brain. Repetition renders these traces permanent, turning flukes into habits that gradually shape our personality because they result in synaptic connections in the brain that make them permanent (Doidge 2007).

Starting from the outside, a word on each of the 4 dimensions. It is followed by an outline of their respective interactions. Once we have looked at the complex ongoing dynamic that forms and dissolves WHO we personally are, we look at the collective sphere (next subsection). Within this complex multidimensional kaleidoscope of our being, every part is complementary to another one. Since the focus of this book is on artificial intelligence, particular attention is placed on the mental dimension.

1.2.1   The Body—Interface Between the Inner and Outer Realms, Between the Self and Others

The realms of our being are connected and constantly influence each other. Functioning like a membrane, the physical dimension—the body, serves on the one hand to experience the environment, and on the other hand to express ourselves in it. Stimuli flow in both directions (this inside-out/outside-in dynamic is further explored below, 1.4.3/1.4.4).

Our senses serve as messengers to connect our inner and outer environments. Seeing that this makes them a prime target for technology, it is worth distinguishing what is part of our sensorial toolkit.

Following the definition of a sense as 'a group of sensory cells that respond to a specific physical phenomenon, and that corresponds to a particular region of the brain where signals are received and interpreted,' neurologists expand the traditional five senses (sight, smell, taste, touch, audition) to at least nine senses. Besides the traditional set they include: Thermoception—the sense of heat (there is some debate that the sense of cold may be a separate sense); Nociception—the perception of pain; Equilibrioception—the perception of balance; Proprioception—the perception of body awareness (close your eyes and touch your nose, if you managed to do so, your sense of proprioception works).

Eco-psychologists put the number of available senses at 53 (Cohen 1994). Their definition of a sense goes beyond the physiological phenomenon/nerve sensor definition—the resulting list can be broken down into 4 categories: radiation senses (e.g., sense of color, sense of moods associated with color, sense of temperature); feeling senses (e.g., sensitivity to gravity, air and wind pressure, and motion); chemical senses (e.g., hormonal, such as pheromones, sense, hunger, thirst); and mental senses (pain, external and internal, mental or spiritual distress, sense of self, including friendship, companionship and power, psychic capacity).

The body reflects our internal circumstances. Conversely, our experience of the environment impacts our internal circumstances, shaping our perspective of the World and hereby our reaction to it. Whatever happens at the center radiates out, like a stone cast into the water. How we experience the physical world sets the overall framework of our mood and mindset and deriving from them our expressions. How we perceive our surroundings based on sensorial triggers leads us to certain interpretations (ideas, thoughts) of the environment that we find ourselves in. These interpretations influence our thoughts and emotions, and thereby impact our behavior.

1.2.2   The Mind—From Mindless to 'Mindfull' to Mindful

Thoughts result from an intricate mixture of genetic disposition, education, beliefs, memories, upbringing, and environment. They influence our emotions and aspirations, our physical experiences and expressions; and are influenced by them (Kahneman 2011). "The heart has its reasons, which reason knows not of" (Pascal in Graham 1998).

Thought processes offer an entry-point to influence our emotions systematically. Because even though we cannot control the environment that triggers our emotional response, nor the emotions themselves, the mental processing that leads from one to the other is a connection that can be systematically addressed. Our present-day perception of a situation derives from past experiences. Our reaction to it is based on a model that is constantly running the risk to be outdated. Our physiological set-up is conducive to maintaining the existing modus operandi. Once a behavior pattern is established, it is challenging to change it because the software and hardware are shaped to keep it running. Keeping an object in motion takes less energy than initiating a new movement or to change direction.

Paradoxically, this same set-up is also our best ally when it comes to the introduction of new behavior patterns. Because the brain can change its own structure and function in response to actual and mental experiences, due to neuroplasticity (Doidge 2007). In that sense, life in a constantly changing environment is the perfect training ground to keep our brain fit. Uncertainty, unpleasant as it is, is conducive to our mental resiliency provided we can manage the deriving stress in a balanced manner (Peters et al. 2014).

Adopting new behavioral patterns does result not only from conscious efforts, but also from observation with the help of mirror neurons. These are a type of brain cell that responds equally when we perform an action and when we witness someone else perform the same action. This neural mechanism is involuntary and automatic, helping the individual make sense of others' behavior based on their own previous experience. Thus, mirror neurons are important assets to our ability to function in society, offering hints to understand both the actions and the intentions of others (Winerman 2005).[3] The challenge is to begin, because neural mirroring only operates if the person who observes has personally performed the action at least once (Johnson 2019).

Whether we perceive something as good or bad, to be liked or despised, results from the interpretations of sensorial inputs in our mind.[4] These interpretations and the deriving judgment are based on experience and knowledge. For example, we might see a large seemingly abandoned bag at the airport; based on experience or too many movies, we jump to the conclusion (described by Kahneman 2011, as System 1 "thinking") that it must be a disguised bomb; a frightening thought which triggers our heart to beat faster, we perspire and quicken our step, driven by a feeling of fear. A more analytical take may persuade us to look closer and hereby discover that the owner of the bag just went to the bathroom; that rational thought, generated by System 2 (Kahneman 2011), disarms our internal flight automatism. Like animals, we have evolved to decide fast with little information, as this is often a critical factor of survival. In an environment that is far more complex than that of our forefathers and foremothers, this leads to mixed results.

1.2.2.1   Ir-Rationality: Thoughts, Errors, Biases, and Decisions

Influence over our action, and that of others, starts by acknowledging and accepting the fact that our choices are not purely rational. Understanding the interplay of brain and behavior, of mind and matter, is the first step to conscious influence over the drivers of our action.

We like to *think* that our decisions are anchored in the intellect, yet science proves that our behavior is determined as much by emotions and physiological impulses as by thoughts (Ambady and Gray 2002). To influence decision-making processes, we must not only think about *what* we feed the brain, but also refine our awareness of *how* the different parts of the brain respond. Simply said—meta-thinking (thought about thought) is required to improve the outcomes of our thinking. If we are alert to errors in the thought process itself, we reduce the risk of ir-rational choices, although we may not be immune to "emotional corruption" (Damasio 2012).

We judge the present based on the past, which means we apply a sometimes-outdated situation to a fresh context. To operate quickly, our cognitive system relies on inherent tweaks. These come on top of the physiological and emotional interferences. They affect our decision-making (perceived desirability of options), judgment (likeliness or causality), memory (e.g., consistency bias—remembering one's past attitudes and behavior as being similar to one's present attitudes), and motivation (e.g., desire for a positive self-image and avoidance of cognitive dissonance) (Bless et al. 2004). Though they get stronger as a result of personal experiences, these tweaks—commonly referred to as bias and heuristics—are largely the same for all human beings in a society. (Thaler and Sunstein 2008).

Heuristics are mental shortcuts to automate, and thereby accelerate decision-making, a necessary survival strategy in view of the limited capacities of the human brain, which needs to cope with and react to an abundant number of external stimuli. A heuristic that consistently leads to an erroneous decision is called a bias (Sunstein 2002). *Bias* is the disproportionate weight given in favor of or against an idea, person, or thing. It may be innate or learned.[5]

Although heuristics are useful shortcuts for everyday judgment calls, they can lead people to make decisions which are not only fast but incorrect, because as situations become more complicated, shortcuts based on simplification are prone to fail. An automated use of these shortcuts may preempt analytical thinking in situations where a more logical process might yield better results (Newkirk 2014).

Cognitive biases are mistakes of the (well-intentioned) human mind. They not only prevent us from perceiving our circumstances objectively, but also from understanding the reality that surrounds us; worse, they significantly influence the way we relate to others (Tversky and Kahneman 1982). These mental filters hamper EXPOSURE to reality as it is. An explicit, conscious mental interrogation of one's own decision-making process, one's attitudes, and one's perceptions would be an antidote. However, using a thorough scientific method to constantly weigh and validate our intuitive automated judgments requires conscious effort and humility (Stallone 2018). As we will see in the next chapters, technology may facilitate that process. Understanding how our brain operates makes it possible to map and influence decision-making processes systematically. Awareness to inbuilt defaults enables us to overcome them and, consequently, change. The point of departure for any endeavor in this direction is to admit that not just the minds of others, but also our own minds, are prone to mistakes. The subsequent step is to understand how these mistakes happen and how they feed our overall operating model.

Since Plato, the decision-making process was defined as either rational or emotional. However, neuroscience increasingly proves that our best decisions result from a blend of feeling and reasoning, depending on the situation (Lehrer 2010).

1.2.3   The Heart—Harbor of Emotions

"Emotions are any mental experience with high intensity and high hedonic content (pleasure/displeasure)" (Cabanac 2002). Various scientific explanations regarding the nature of emotions exist. According to 'appraisal theories,' emotions are part of an adaptive, flexible set of responses to the environment. Human emotional behavior is highly dependent on individual, social, and historical factors that come into play when individuals assess their environment. The 'cognitive appraisal theory' posits that emotions are judgments about the extent to which your current situation meets your goals (Watson and Spence 2007). Others have argued that emotions are perceptions of changes in the body such as heart rate, breathing rate, perspiration levels, and hormone levels. In this view, happiness, sadness, and anger are different kinds of *physiological perceptions*; rather than judgments, they are mental reactions to different kinds of physiological states. Cognitive appraisal and physiological perception can be combined into a unified account of emotions (Thagard and Aubie 2008), with the brain acting as a multitasking parallel processor of the body and the surrounding environment.

Emotions can be categorized in many ways. One common way is the distinction in seven basic emotions (Joy, Surprise, Sadness, Anger, Fear, Disgust, and Contempt) proposed by Ekman and Friesen (1971). Each one of these basic emotions has several sub-emotions beneath it (i.e., Joy covers Delight, Ecstasy, Awe, Satisfaction, Gratitude, Thrill, Pleasure, etc.). Another classification is the settlement of emotions along axes that connect apparent opposites (pleasant-unpleasant, tension/excitation-relaxation/calm) (Kragel and LaBar 2016). Each emotion comes in varying degrees of intensity. When emotions reach a state of arousal, they drive us into action. The connection between the body and our emotions is tight as the latter derive from previous experiences, inherent values and beliefs, built-in fallacies of the brain, as well as physiological factors such as hormones and neurotransmitters (Bechara et al. 1999).

Many emotions are triggered by autopilot thinking. The resulting reaction is instinctive rather than rational. However, due to the fluid interplay of our hardware (body including the brain) and software (emotions, thoughts), emotions, like thoughts, are not immutable. One follows the other and is subject to conscious influence. Becoming aware of the underpinning chain reaction that connects the following pairs makes it possible to snap out of our inherent autopilot.

1.2.3.1   Thought/Bias > Bias/Emotion > Emotion/Decision > Decision/Behavior

Systematically analyzing one's emotional state and its manifestation through speech or behavior involves a 4-step process:

1. Acknowledgment of the emotion, by consciously labeling it;
2. Appreciation of the values that underpin the emotion/thoughts triggered by these values in that situation;
3. Acknowledgment of the consequences of the intuitive reaction that we are about to engage in; and
4. Appreciation of ourselves and the action that we are about to take.

This analysis involves the cognitive as well as the emotional level. Activating the prefrontal cortex (hardware) where the decisions are shaped has been shown to suppress emotional reactions (software) (Korb 2015) [Hardware influences software]. Thus, consciously analyzing whether our intended action will have positive/negative/neutral consequences for others can stop instinctive emotional reactions such as fear, anger, or greed,[6] which influences the manifestation of our emotions in the form of behavior [Software influences hardware]. Reappraising a situation changes the way the situation is perceived and reacted upon by individuals (Rock 2020). Repeated regularly, this analytical lens eventually creates a new habit—which is reflected in synaptic connections in the brain (Doidge 2007). It is like an analytical self-help model to foster Emotional Intelligence (Goleman 1998). These models matter when it comes to the design of empathic AI. To be conducive to human emotions, the latter must be understood first.

1.2.3.2   Empathy Versus Compassion

When humans and machines merge, as it will be explored in the ensuing chapters, empathy and compassion matter more than ever. Representing one of the last strongholds of humanity, they are required to achieve a society that is marked by solidarity (Stead and Garner Stead 1994). Unless individuals feel *with* those in need, they do not sustain their efforts of support over a longer period or until the suffering is relieved.

Empathy is the ability to sense other people's emotions, coupled with the ability to imagine what someone else might be thinking or feeling (Merriam Webster 2018). Researchers often differentiate between two types of empathy. 'Affective empathy' refers to the sensations and feelings we get in response to others' emotions; this can include mirroring what that person is feeling, or just feeling stressed when we detect another's fear or anxiety. 'Cognitive empathy,' sometimes called 'perspective taking,' refers to our ability to identify and understand other peoples' emotions. Both are part of our evolutionary history and have deep roots in the human brain and body (Waal 2005). Elementary forms of empathy have been observed in primates, in dogs, and in rats. Empathy is associated with two different pathways in the brain, and scientists have speculated that some aspects of empathy can be traced to brain cells that fire when we observe someone else perform an action, in much the same way that they would if we performed that action ourselves (Marsh 2012). The same 'mirror neurons' mentioned above that are at play in the process of acquiring new skills and habits maybe involved in the activation of empathy. Empathy is a step toward understanding the needs of others and feeling the consequences of these (unsatisfied) needs, yet it does not necessarily result in the desire to help someone in need.

Compassion, defined as "concern for the sufferings or misfortunes of others" literally means 'to suffer together' (Goetz et al. 2010). It can be defined as the feeling that arises when you are confronted with another's suffering and feel motivated to relieve that suffering (Strauss et al. 2016). Cast as benevolence, David Hume described it as the prime way to relieve social ailments: "Benevolence offers the merit of meeting human need and bestowing happiness, bringing harmony within families, the mutual support of friends, and order to society" (Hume 1740).

Emotions of any kind can be assessed and amended. Even though the genetic set-up plays a role, it is less determinant for positive emotions than for negative emotions[7]. Thus, the degree to which a person experiences Compassion is changeable.[8] Research has uncovered evidence for a genetic basis to empathy, while suggesting that people can enhance (or restrict) their natural empathic abilities (Marsh 2009). It is a trait which parents can nurture from an early age (Borba 2016). Generosity, the physical manifestation of compassion, falls in the same category. Defined as selfless action, it does not always result from compassion, but may be the consequence of a cerebral analysis of one's own responsibility. On the other hand, while genuine compassion always involves an authentic desire to help (Seppala 2013), this desire may not always find a physical expression. Awareness of our innate ability for compassion prepares the ground for generous behavior. Retraining our perspective from negative to positive, from danger to drive, from comfort-zone to creative compassion is not only possible but facilitated by nature and necessary to make the best out of the range of opportunities placed in our lap by technology (see Chapter 2).

1.2.4   The Soul—Essence of Who We Are

Embodying our aspirations, the soul encompasses the universal human quest for meaning in life. Whatever we feel, think and do, is rooted in this craving for a purpose that transcends the self (Frankl 1985). "Your soul is that part of you that yearns for meaning in life, and which seeks for something beyond this life" (Graham 1998).

Anchored in the soul are values. They are cornerstone of our belief systems, which in turn underpin our aspirations and can serve as an inner moral compass to guide our decision-making. Defined as broad moral preferences concerning appropriate courses of actions, they reflect a person's sense of right and wrong or what *ought* to be. They influence (both sub-consciously and consciously) our behavior, impacting our aspirations, emotions, and thoughts. Serving as a reference point, they facilitate our decision-making processes, with the possibility of comparative ranking (Seligman 1991).

1.2.4.1   Values The Anchor of Aspirations

Aristotle believed that a set of core values should be manifest in the behavior of all human beings. These were courage, honesty, friendliness, wittiness, rationality in judgment, mutually beneficial friendships, and the pursuit of knowledge and truth. In his vision, two kinds of virtue exist: intellectual and moral. The former derives from instruction, and the latter comes by habit and constant practice. Like a musician learns to play an instrument, we learn virtue by practicing, not by thinking about it. Ideally, we live our lives based on our core values, using them subconsciously as the foundation from where we make our decisions. Values are anchored in our intrinsic moral foundation (Ross and Brown 2009).

Indeed, developmental psychology has shown that we are born with an innate value 'draft' of the world. "The initial organization of the brain does not depend that much on experience. Nature provides a first draft, which experience then revises. 'Built-in' doesn't mean unmalleable; it means organized in advance of experience" (Marcus 2005). Across continents and cultures, there is commonality on 'five foundations of morality' (Haidt 2006):

1. Purity/sanctity—The idea of virtue comes from controlling what you do with your body. It refers to the purposeful orientation of our being toward something that is bigger than yourself.
2. Harm/care—Humans are mammals whose neural and hormonal programming renders them prone to bond with others, to care for them. We have evolved to feel compassion for others, especially those who appear as weak and vulnerable, which in turn gives us strong feelings about those who cause harm.
3. In-group liaison/loyalty—Humans are unique in their ability to cooperate and form groups among those that share no filial links (Nowak and Highfield 2011). We are social beings by birth.
4. Authority/respect—It is not only based on power and (threats of) violence but also voluntary deference and love.
5. Fairness/reciprocity—It is illustrated by "The Golden Rule" which has found entry in belief systems through times and cultures, making it a central part of many religions.[9] By translating the 4 other values into action, it encompasses them all.

From our values derives our drive for purpose in life. The WHY of being alive. Awareness to the values that underpin what we want equips us with a set of easily accessible decision-making parameters. These make it easier to align our aspirations and our actions. As seen earlier, repeating this type of behavior makes it easier over time because the emerging synaptic connections facilitate it. The 4 central values, on which we will concentrate on here, are: Generosity, Compassion, Honesty, and Courage.[10] They are rooted in the 4 dimensions of our being (soul, heart, mind, body) encompassing the behavioral, intellectual, emotional, and aspirational features of experiences and expressions. The fifth one—Fairness/Reciprocity (the Golden Rule)—conditions collective wellbeing (and is discussed in the following sub-section).

- Generosity of the soul is mirrored as kindness in the interaction with others. We aspire to be and to be perceived as generous. The *Awareness* and pursuit of a purpose whereby we dedicate our time and energy to a cause that goes beyond our own immediate interests reflect it. Linking this to the moral foundations seen above, Generosity is rooted in the Purity part of our moral draft, involving selflessness and the quest for something larger than our own individual benefit.

- Compassion is the feeling of not merely suffering with others, but to genuinely desire to relieve that pain. It is conditioned by the level of compassion that we experience for ourselves and our ability of *Acceptance* of our own limitations. It links to the Care/Harm part of our common foundation.

- Honesty is an intellectual effort as it entails to consciously overcome the urge of hiding what we dislike or feel ashamed of. Honesty with ourselves is essential for personal growth, because the *Alignment* of our aspirations and our actions passes via the acknowledgment of behavior that is presently out of line. Self-delusion is time- and energy consuming. Acknowledging the status quo frees energy spent on pretending, to change the status quo. This links to the Loyalty component of our moral draft, because dishonesty erodes and eventually breaks down the trust needed for loyalty and genuine interpersonal relations.

- Courage manifests the shift from aspiration to action, illustrating the ability to not only believe in something, but stand up for it. *Accountability* for what we do is rooted in the Respect part of our universal moral draft. It has found entry in philosophies across cultures and generations (Mullin 1983). From the three Rs of the Dalai Lama (respect for us, respect for others, responsibility for everything we do) to corporate mission statements and State declarations, it is the leadership mantra that is on everyone's lips, while slipping away quietly from the radar of behavior (Ellis 2016).

Values shape aspirations which influence our environment—from relationships to behavior, from choices to personal identity. Our choices influence which values we express more and which ones we numb. When our actions and words are aligned with our aspirations and values, we feel coherent, experiencing inner peace. Cognitive dissonance ensues when our experiences and expressions are in contrast to our internal value system (Festinger 1962).[11] When we continuously ignore our instinct to pursue a certain value, an undefinable inner discomfort arises. Internal misalignment expands gradually, affecting our mental and physiological wellbeing and our relationships. It influences how we perceive others and how they perceive us. Manifested as inauthentic appearance it is a killer of healthy psychological functioning, subjective wellbeing (Goldman and Kernis 2002), and social influence (Strohminger et al. 2017).

1.2.4.2   The Golden Rule—A Golden Standard

Thinking about Soul entails often a link to religion. As stated earlier, this is not the case in the context of POZE. Rather, it relates to the universal spiritual fiber that transcends everyday needs and wants. "We are not human beings having a spiritual experience. We are spiritual beings having a human experience" (de Chardin 1955). It manifests in our quest of a purpose, a WHY. The question what we want to do with our lives is a central part of human evolution. One of its expressions is reciprocity, or the ability and desire of doing for others what we perceive as desirable for ourselves. Living up to the inherent aspiration of WHY is nurtured by our ability for Compassion (of feeling with others and behaving accordingly). Thus, the aspiration for Generosity is nurtured by the emotion of Compassion, and both find a reflection in the logic of reciprocity.

Whether phrased as "Do not do to others what you would not like them to do to you" or "Do to others what you wish done to yourself," the Golden Rule has found entry in societies around the world. It translates our moral blueprint into action. Propounded by Confucius five centuries before Christ, it was believed that obeying the Golden Rule would bring those who practice it to the transcendent value of "human-heartedness."[12] Though it is treated differently across religions, an ethic of reciprocity can be found in all of them (Flewd 1979).[13]

It is a prominent concept in Buddhism, Christianity, Hinduism, Islam, Judaism, Jainism, Sikhism, Taoism, Zoroastrianism, and many others (Blackburn 2001). In 1993, 143 leaders of the world's major faiths endorsed the Golden Rule as part of the "Declaration Toward a Global Ethic" (Parliament of the World's Religions 1993).

Beyond the moral imperative, it is in our personal interest to enact Kant's 'categorical imperative.' ("Act according to those maxims that you could will to be universal law.") Because where internal and external valuation, intrinsic and extrinsic values clash emotional, mental and physical discomfort arises. Remaining aligned to one's internal compass is conducive to inner peace, and social harmony. It offers stability, whereas the blind adjustment to a prevailing social codex that stands in discrepancy with one's inner belief system results in cognitive dissonance (Gino et al. 2015), which is detrimental for oneself and others as seen above. Technology cannot replace values yet it can keep them present on our inner radar. We know and feel what is 'right,' yet in the heat of a situation we may act against that inner judgment because emotions and bias snap in. Similar to a fitness tracker, a value wearable could keep us on track by tracing deviations from our moral map. We will explore this option further below (Sect. 4.2).

****

Before moving on to the collective dynamics a word on interplays is useful to tie up loose ends—.

1.2.4.3   Mind and Matter

There is not one neurological network for each emotion or even for the experience of emotion versus cognition (Oosterwijk et al. 2012).[14] They are all part of the same network, a living spiderweb that connects every aspect of our inner and outer experience of that state called reality (Barrett 2006). The brain perceives our emotional and mental states through situated conceptualizations which combine three sources of stimulation: The first is sensorial stimulation from the world outside the body. Exteroceptive sensory signals come from light, vibrations, chemicals, sound, etc. The second type of stimulation comes from interoceptive sensory signals within the body (the internal milieu, which includes the microbiome that resides in the gut and has a direct connection to the brain). The third source of stimulation is prior experience, also referred to as memory or category knowledge, that the brain makes available in part by the re-activation of sensory and motor neurons. These three sources—sensations from the world, sensations from the body, and prior experience—are continually available. They influence how we experience the environment and express ourselves in it. These influences impact how we perceive and manifest the following pairs.

1.2.4.4   Purpose and Passion

Purpose and Passion are not synonymous. Both are inherent to humankind. Passion is personal and primal. It is a feeling that is anchored in the body (e.g., sex drive) and the heart. Pursuing it is an emotional motivation. On the other hand, the type of Purpose we veer toward is influenced by our personal, unique mix of characteristics which derive from our background, upbringing, physiological, and mental assets (mind, not just heart). Passion and purpose may not always come together. Ideally one nurtures the other[15]; with passion as an inject of desire and pleasure, and aligned with purpose as the guidance to translate beliefs into practice. Purpose often entails a contribution to the wellbeing of other people; thus, if it is present yet bare of passion, the feeling of compassion toward those others may establish the missing emotional link. Heart and Soul unite.

1.2.4.5   Happiness and Meaning

The pair Passion and Purpose is related to the duo Happiness and Meaning. The latter correlate and often feed off each other. The more meaning we find in life, the happier we typically feel, and the happier we feel, the more likely we feel encouraged to pursue even greater meaning and purpose. However, it appears that happiness has more to do with the satisfaction of our own needs, getting what we want, and feeling good, whereas meaning is more related to uniquely human activities such as developing a personal identity, expressing the self, and consciously integrating one's past, present, and future experiences (Baumeister 1992). There is general consensus that meaning has at least two major components: the cognitive processing component involves making sense and integrating experiences, and a purpose component, which is more motivational and involves actively pursuing long-term goals that reflect one's identity and transcends narrow self-interests.

Across cultures and generations, purpose was found to be a parameter of happiness. Identifying and pursuing a purpose is a necessary condition of fulfillment, which is the quintessence of happiness, but also of physical health and longevity (Seligman 2011). Progress toward purpose is not marked by peaks but peace. The road is the goal. Even though meaning-making may be associated with negative emotions in the moment, it contributes to greater resiliency and wellbeing in the longer-term (Abe 2016).

An intent-less chain of pursuits is inherent to the operating system of nature (Gould 1989). A bee does not visit different flowers in order to collect and spread their pollen as part of the natural process that leads to fruits; it does so to collect honey. The result of pollinated trees that bear fruits still ensues. As humans, we can identify both our purpose and the indirect (or unintended) impact of our pursuits. "Humans may resemble many other creatures in their striving for happiness, but the quest for meaning is a key part of what makes us human, and uniquely so" (Baumeister et al. 2013). We should make use of that ability within a holistic perspective.

1.3   SOCIETY

Nothing occurs in a vacuum. Like the 4 internal dimensions that shape an individual's being, society is influenced by multiple dimensions. The latter evolve along with the individuals that constitute that society at a given point in time. It is a constantly shifting dynamic. Influenced by mutual social interrelations, this living social kaleidoscope is referred to as the mmmm-matrix in this book (Fig. 1.2).

- The **micro**-level denotes on the one hand the 4-dimensional internal composition of every individual's being; on the other hand, it refers to the role of the individual in society.
- Every person is at the same time part of various **meso**-entities, different types of communities/institutions/organizations, be it a family, church, workplace, school, political party, or sports-club.
- These meso-entities function as an intermediary between the person and the subsequent **macro**-level, which encompasses the economic, political, and cultural spheres that we evolve in, such as a country.
- Respectively and combined, these three dimensions form part of the **meta**-dimension[16] which includes also non-anthropocentric dynamics that influence our life—Nature—as well as supra-national entities such as the United Nations, due to their global mandate and impact (Fig. 1.2).

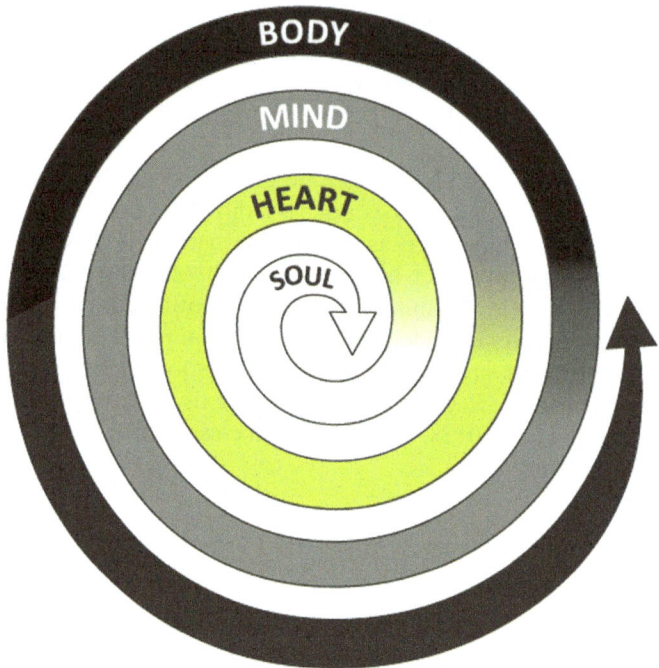

Fig. 1.1 Spiral Soul to Body. Everything is connected, from the inside out and from the outside in. Our soul finds its expression in our aspirations. These influence the heart, which is the source of our emotions. How we feel influences our mind. How and what we think influences our overall wellbeing, and our behavior, from the inside out and from the outside in. The body is the interface between our inner and outer realities. Experiences influence our mind and heart, our thoughts and feelings. Physical experiences influence our inner realm, while the latter shapes our expressions, our attitude and behavior in the outside world

Every meso-entity (from families and schools to firms, parties and universities, from public service entities to multi-billion conglomerates) is constituted of myriads of micro-entities (individuals), and depending on their size and nature, it may entail multiple meso-entities.[17] Each meso-entity operates along the same principle of interconnected dimensions as does the smallest unit in its composition—the human being. How an

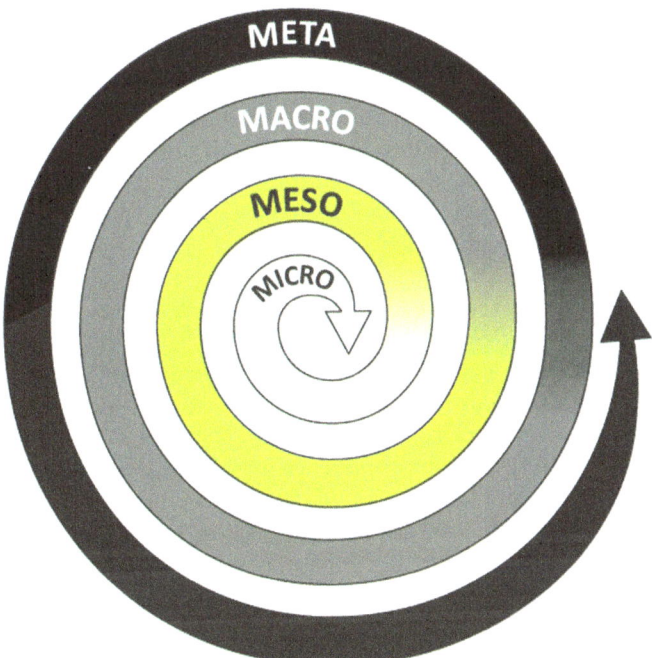

Fig. 1.2 Spiral Micro to Meta (The mmmm-matrix). Everything is connected, from the inside out and from the outside in. Individuals represent the micro dimension of a multi-dimensional system. They form, shape and experience the meso level which results from the communities they are part of (voluntarily or biologically, directly or indirectly). The contours of economic, political, cultural systems form the macro level. Micro, meso and macro dimensions operate within a meta system that entails the three other dimensions, Nature and supranational entities such as the United Nations. Within a seamless continuum one dimension is influenced by the others and influences them in turn

entity evolves is thus determined by both the expression of the particles that constitute it and the interplay between them.

Supranational (meta) and national (macro) dynamics are both, the cause and consequence of the constant interplay that occurs at individual (micro) and institutional (meso) levels. Seeking change in one dimension must necessarily consider how one level is affected by the others and how it affects them.

Every individual is, consciously or not, willingly or not, simultaneously part of several meso- and macro-dimensions and the overall meta-dimension. Thus, the core of any entity, and of our collective being overall, are individuals. For example, an institution is made of individuals; thus, its institutional identity (internal) and organizational operation (external) derive from the identity and behavior of the people who work in it. Individual existences are marked by aspirations, emotions, thoughts, and sensations. Since individuals (micro) and sub-national institutions (meso) shape the macro-dimension, and are major components of the meta-dimension, the future of society hinges on the course of these two spheres of life.

Individual life and collective life are intertwined. Our connective and communicative environments are embedded with our bio-psycho-social influences. An "interpersonal neurobiology" (Siegel in Kingsley 2011) arises as our social relationships influence and affect our nervous system. Since the neurological functioning of the human being is not separate from the environment, our social and cultural influences and experiences are rewiring our neural circuitry. As a consequence, we need positive and constructive external stimuli to foster and support conscious development. "I believe that all education proceeds by the participation of the individual in the social consciousness of the race. This process begins unconsciously almost at birth, and is continuously shaping the individual's powers, saturating his ideas, and arousing his feelings and emotions. Through this unconscious education the individual gradually comes to share in the intellectual and moral resources which humanity has succeeded in getting together. He becomes an inheritor of the funded capital of civilization" (Dewey 1897). Humans shape humanity and vice versa society influences those who constitute it. Technology, given its influence on our behavior and the deriving social dynamics may serve the common good or impede it.

Optimizing our influence on others (see below) begins by understanding our own operating model first. Self-awareness is necessary to understand others, because they operate based on the same basic human 4-dimensional model as we do—though the result is as unique as each person's past and present. Optimizing our interaction as a society, the topic of the next chapter, entails the synchronization of 32 billion individual dimensions (8 billion people × 4 individual dimensions) and is compounded by the collective dynamics that ensue at the meso-, macro-, and meta-levels due to global influences.

Social harmony radiates from the center of this multidimensional kaleidoscope. It is nurtured and blocked by the aspiration of individuals to leave (or not) a meaningful trace. The harmony of the collective is hijacked when individual components are out of synch. When aspiration for purpose that serves the common good is taken over by the passion of personal pursuit that is at the expense of others. Whatever fragmentation occurs within one dimension has repercussions on the others. Appreciating the involved factors equips us with the necessary point of departure to address them; not merely to re-establish inner balance, but to systematically cultivate the synchronization of the multidimensional reality that we evolve in (Figs. 1.3 and 1.4).

The interaction of individuals and institutions shapes major parts of our collective existence. Summarized as the *mmmm-matrix*, the flows at stake can be visualized. To optimize the interplay of micro-, meso-, macro-, and meta-dimensions that shape our life is the big question that humanity is tasked to solve. Bringing the ongoing dynamics within one coherent, organically evolving matrix helps us map and analyze these gazillions of interactions overtime. Used with the intention of social optimization not commercial interest, it is a gigantic endeavor and an excellent use of technology. Provided it is pursued with awareness to the flows of influence at stake.

1.4   INFLUENCE

Understanding ourselves helps us to understand others. Since we all operate with the same basic human set-up, we can distill certain approaches that resonate with all of us. Rather than starting with features that separates us, the present PERSPECTIVE offers a common denominator to build influence, and bridges.

1.4.1   The Scale of Influence

Influence on others involves the twice 4 dimensions that underpin our experiences, expressions, and environment. To influence someone in view of sustainable behavior changes, we must use a 360-degree approach. Unfolding along the *Scale of influence*, sensorial, intellectual, emotional, and aspirational components are complementary, occurring simultaneously or subsequently[18] (Fig. 1.5) (Walther 2020b). A word on each.

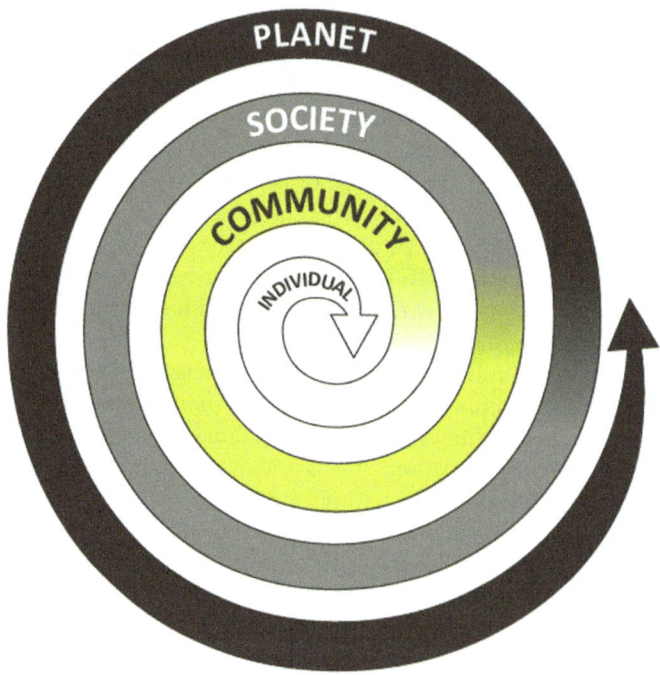

Fig. 1.3 Spiral Individual to Planet. In the perspective of the present prism individuals are at the same time a unit that forms part of a whole, and a 4-dimensional entity that is ruled by the same interconnected dynamics that determine the collectivity which it is part of. Engaged in an ongoing spiral dynamic, from the center to the periphery and from the periphery inwards, multiple dimensions continuously interact and influence each other. This internal two-way interaction influences who and how we are, and what we do as part of our environment

The first component is *Inspiration*. Anchored in the soul, our aspiration for meaning is triggered when we witness individuals who perform the type of action that illustrates the values that we aspire to. Lasting social transformation starts here, with individuals who witness others do what they would do if they could get over the draw of psychological inertia, to manifest their inherent values, which would be beneficial to their own wellbeing (Kuppens et al. 2010)[19] and for that of others.

The second component is *Inducement*. It relates to the heart, our emotions. Emotions are a central part of the decision-making process.

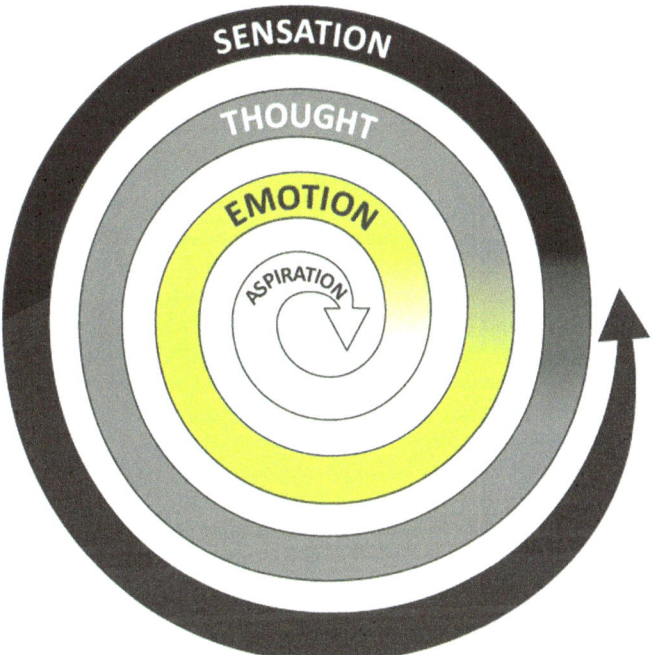

Fig. 1.4 Spiral Aspiration to Sensation. Every individual is a micro illustration of the multi-dimensional dynamics that shape the 'reality' that shapes them. We occupy an organically evolving kaleidoscope that is constantly changing, from the inside out and from the periphery to the center. What we want (Aspiration) influences how we feel (Emotion) which influences what we think (Thought) and impacts not only how we experience our environment (Sensation) but how we express ourselves in it, which influences the situation we are in. Everything is connected. From the inside out and vice-versa

When we care about an issue, the shift from awareness to action is more likely to happen (Frijda 2004). Knowledge without emotion is useful but sterile when it comes to social transformation. Individuals dare to take risks if they care about something/someone.

The third component is *Intrigue*. The mind seeks to make sense of the world it evolves in. Arguments must combine obvious facts and more subtle facets to not only address the need for rationality, but also tickle the human curiosity to learn more. Since the mind is constantly exposed

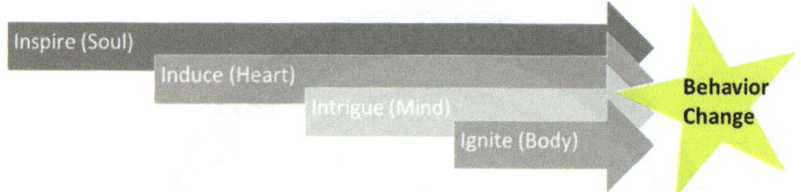

Fig. 1.5 Scale of influence. Genuine influence involves the 4 dimensions of (at least) two people—of the one who influences and of the one who is influenced. Genuine influence is exerted by people who (1) Inspire purpose—because their own aspirations and actions are aligned; (2) Induce feelings—because their audience starts to relate to the issue at the emotional level which is needed to trigger the shift from acknowledgement to action; (3) Intrigue interest—because now the curious mind wants to learn and understand better; and (4) Ignite behavior change—because physical action reflects the transition from inner to outer realm. Sustainable influence requires inspiration which ensues when the aspiration for meaning is touched in a human being. This intensity is conditioned by inner alignment, of aspiration and action; words and works. It is an inside-out/outside-in dynamic that mutually propels itself (see Sects. 1.4.3 and 1.4.4)

to various stimuli, external and internal, surprise is an asset. Persuasion requires lively details and coherent facts in a balanced package that appeals to the intellectual and the emotional dimension.

The 4th component is *Ignition*. Action is the physical manifestation of our aspirations, emotions, and thoughts. It is the counter-piece to Inspiration because it is the external embodiment that carries our values into the world, making them tangible to others. Through this external manifestation, a new Scale of Influence may be triggered, if the behavior of the influenced individual is witnessed by another person, and the latter feels compelled to change their own course of action.

1.4.2   Influencing Mindsets

We influence others with varying degrees of intensity. Our external influence depends on the extent of influence that we have on ourselves. Each component of the *Scale of influence* is conditioned by the 4 outcomes that are looked at in this book, and conditions them in return. A word on each stage.

Awareness—The first stage is the understanding of our internal setup and the interpersonal interactions (micro, meso, macro, meta) that derive from it. The interactions that we have with others are only to a limited extent under our control. Amid the prevailing uncertainty that caracterizes life and living, a holistic PERSPECTIVE of the dynamics that influence ourselves, internally and externally, improves our ability to optimize our own life, which in turn not only impacts the effect of our presence/absence on others but equips us to systematically influence the multi-dimensional social continuum that we are part of.

Knowledge of the dynamics at play allows us to harness our natural and artificial equipment, hardware and software, inherent and external, to maximize the type of influence we want.

Acceptance—The second stage is to accept the conditions that shape who we are. It is a choice. The alternative is to deny what influences us and others, and the status quo that derives from it. We cannot make unknown information that has settled into our mental sphere, but we can decide to ignore it. Once we know and accept the circumstances and the causes and consequences that relate to them, we can use our understanding to shape that influence systematically. OPTIMIZATION internally and toward others ensues. Once we accept what is, we can 'influence the influence,' shifting from passive to active, from being influenced to influencing. Once we grasp the streams of influence that we are exposed to 24/7, we can protect ourselves (at least partially), while at the same time strengthening our own influence on ourselves and others. Certain technological tools are prone to influence us. Used with a passive consumer attitude they lull us in a virtual comfort-zone that is shaped by others. Understood and used well, they can increase our influence internally and externally.

Acceptance is a choice and an attitude that conditions action. Based on *Awareness* of the implications, we must think and decide, individually and collectively, WHY we want technology to be part of our life.

Time is a continuum of past, present, and future. Who we are now is influenced by past action; and what we are doing now influences Who we will be in the future. The choices that we make today about our relationship with technology determine the future. We must accept this link and its consequences. *Alignment*—The third stage is conditioned by the two others. It entails conscious influence on the parameters that we have acknowledged and accepted. Once these two hurdles are overcome, we can systematically pursue the coherence of those parameters, including their mutual influence onto each other, on our perception of 'reality' and on the environment in which we exist. This is in our interest and in the interest of others.

Internal alignment conditions the level of our influence on others, because it is accompanied by the gradual harmonization of our external interactions. When our aspirations, emotions, thoughts, and behavior are aligned, and anchored in human values, we are perceived as authentic and thus trustworthy. Eventually, we reach the ZENITH of our being, individually and collectively—because pro-social action is no longer a conscious effort but an intuitive (re) action to the system that we evolve in.

As pointed out in the Introduction—*Alignment* is both a process and an outcome. It relates to the ability of consciously manifesting a harmonious ensemble of our aspirations, emotions, and thoughts. Herein 'the journey is the goal.' We may not reach the ZENITH of our own best self, the fulfillment of our inherent potential, but the conscious effort of pursuing this goal, is useful by itself. Because along the journey we learn and improve. As our behavior reflects our values ever more often our wellbeing improves; inner harmony ensues; simultaneously, our external interactions improve because they are characterized by generosity, compassion, honesty, and courage. Harmony in our environment ensues—which in turn contributes to our personal wellbeing.

Accountability—Stage 4 reframes the potential of individuals to bring about change in terms of responsibility. Being aware of our latent power and not using it is a choice with consequences. Because acknowledging a problem and not addressing it makes us part of that problem. We are accountable to ourselves and toward others for the passive and active influence that derives from our resources, innate and external. The more we have, the more we can give. The 4th stage relates to the responsibility of living our values. Generosity, compassion, honesty, and courage are the minimum moral standard to uphold. Whether we pursue and uphold them is under our immediate control; it influences our interactions and thereby others. We are accountable regarding the use that we make of that influence, and the lack thereof.

1.4.3 The Inside-Out and Outside-in Principles

Nothing happens in vacuum. Every human being is a micro-illustration of life and the same laws apply everywhere (Sect. 1.1). Change starts from the inside out and is nurtured from the outside in. A shift in the prevailing dynamics can be triggered from either side. In the following, we look at the two-directional dynamic that influences our personal and collective sphere.

The prism of the *mmmm-matrix* (Fig. 1.2) turns the power of influence into a concept that is at the reach of everyone. WHAT we do and WHY we do it influences WHO we are and WHERE we go—and inversely. Based on a perspective of indivisibility, the present paradigm offers two complementary approaches that operate within a spiral dynamic that turns in both directions simultaneously. The first approach concentrates on social transformation by pursuing the inside-out principle (IOP), working on the aspirations, emotions, and thoughts of individuals to influence their experience of the environment, and thus their expressions (word/action) in it. The second approach is based on the outside-in principle (OIP). It looks for the modification of behavior as a way of inducing new experiences that trigger new types of emotions, thoughts, and hereby a new set of neurological pathways to eventually facilitate the transition from effort to habit.

a. Change from the inside out

The brain is a pattern-recognition machine. Establishing automatisms is the only way for the brain to process the huge quantity of information that we are relentlessly subjected to via our senses. Starting from an early age, the brain starts forming rules about the world based on observations. As it forms these rules, errors tend to occur (Scherer et al. 2001). As time goes by, we are getting used to these rules and the resulting automatic thoughts until we do not even notice the rules that triggered those thoughts in the first place (Sect. 1.2.2). Awareness to this trigger mechanism which evaluates current experience based on the past is the first step to change our present-day expressions. This is facilitated by our physiological set-up. Hardware serves software.

We can transform our behavior, our habits, and gradually our personality throughout our lifetime. Neuroplasticity, the ability of our brain to change itself, is conducive to change (Doidge 2007). As experience proves the 'new' thought accurate over and again, the brain slowly adopts new rules, while reducing the negative feelings associated with the erroneous or outdated rules. New software overwrites old software. Fresh neural pathways are created, replacing the existing ones. Newer hardware replaces outdated hardware. The essential element is to continue the corrective efforts. Change in mental and physical infrastructure happens gradually, in the same slow incremental way in which the initial setup evolved to become the status quo.

Simultaneously to the inside-out principle (IOP), the second approach focuses on change via the 'outside-in principle' (OIP).

b. Change from the outside in

Our behavior impacts how we perceive ourselves and others. Thus, every action is prone to shape our personality in one way or another. Exposing ourselves on a regular basis to new experiences challenges established habits and beliefs. "Outsight is the fresh, external perspective that comes from doing new and different things – plunging ourselves into new projects and activities, interacting with people outside our daily routines, and experimenting with new ways of getting things done" (Ibarra 2015). The OIP is the complementary opposite of learning by self-reflection, in which we seek insight into our past behaviors.

Furthermore, the environment impacts our behavior; modifying the external setting may be conducive to certain behaviors and making the desired behaviors easier to perform than others. "When we think about sending a rocket to space, we want to do two main things. The first one is to reduce friction. We want to take the rocket and have as little friction as possible so it's the most aerodynamic possible. And the second thing is we want to load as much fuel as possible, to give it the most amount of motivation, energy to do its task. And behavior change is the same thing" (Ariely 2008).

Our mind-matter set-up facilitates the acquisition of new behaviors. We are configured to change and adjust to organically evolving circumstances. "The brain is a far more open system than we ever imagined, and nature has gone very far to help us perceive and take in the world around us. It has given us a brain that survives in a changing world by changing itself" (Doidge 2007). It is also set up for cooperation (Nowak 2011).

1.4.4   Why It Matters—The Common Good

One of the central challenges to building a society that is not ruled by technology but facilitated by it to the inclusive benefit of all is to induce social norms of solidarity. Behavior that does not yield immediate advantages for us may seem counterintuitive. However, as seen earlier, reciprocity is built into our DNA. Compassion and generosity are rooted in our first moral draft and coded into our operating system. We can use technology to prime ourselves to reveal these inbuilt features and systematically engage in pro-social behaviors.

Building on the holistic PERSPECTIVE outlined in this chapter, we proceed to look at ways of using it to OPTIMIZE what influences us and others in view of the common good.

The latter is used in this book as a baseline to orient our individual and collective efforts. It stands in opposition to self-interest, even though both are not mutually exclusive and may even benefit each other. Used in philosophy, economics, and political science with various definitions, "the common good refers to the benefit or interests of all" (Diggs 1973). Most philosophical conceptions of the common good fall into one of two families: substantive and procedural. According to substantive conceptions, the common good is that which is shared by and beneficial to all or most members of a given community, whereas according to procedural formulations, the common good consists of the outcome that is achieved through collective participation in the formation of a shared will (Simms 2011). In the present context, we look at both aspects.

The OPTIMIZATION of individual and collective dynamics serves the common good as process. The result is a transitory ZENITH where Society is optimized for the maximal benefit of all. We look closer at them in the next two chapters.

NOTES

1. As framed in the much quoted dialogue between the Cheshire Cat and Alice: "Alice: Would you tell me, please, which way I ought to go from here?

   The Cheshire Cat: That depends a good deal on where you want to get to.
   Alice: I don't much care where.
   The Cheshire Cat: Then it doesn't much matter which way you go.
   Alice: ...So long as I get somewhere.

The Cheshire Cat: Oh, you're sure to do that, if only you walk long enough" (Carrol 1865).

2. The word soul is used without religious connotation, referring to the immaterial substance of our being.

3. Mirror Neurons could help explain how and why we read other people's minds and feel empathy for them. They were discovered in the early 1990s, when a team of Italian researchers observed how individual neurons in the brains of macaque monkeys fired both when the monkeys grabbed an object and when the monkeys watched another primate grab the same object. "If watching an action and performing that action can activate the same parts of the brain in monkeys--down to a single neuron--then it makes sense that watching an action and performing an action could also elicit the same feelings in people" (Winerman 2005).

4. Decisions are performed by the prefrontal cortex which interacts with the amygdala and insula that process information about physiological states.

5. The most common cognitive biases are the following:

- **Confirmation bias**: whereby we surround ourselves by people and platforms, online and offline, whose statements and opinions confirm our current beliefs.
- **Negative bias**: the tendency to give more weight and attention to negative news than to positive news. For example, although statistics show a reduction of poverty rates compared to the past, people believe that development is stalled or worsening; the case is similar for wars and violence.
- **Bandwagon effect**: the inclination to believe/do certain things because many other people believe/do so. People tend to walk the way of least mental resistance and thinking for yourself means active effort.
- **Illusory correlation**: we may be inclined to assume a cause and effect relationship that does not actually exist.
- **Dunning-Kruger effect**: leads certain people to inadequately assess their level of (in)competence by overestimating their knowledge. Attributed to their low level of knowledge and competence, this lack of awareness deprives them of the ability to critically analyze their performance and accurately perceive low performance level.
- **Overconfidence effect**: leads people to overestimate their own abilities with respect to their own actual performance.
- **Illusory superiority**: when the overconfidence effect is in relation to the inferiority of another's performance.

6. Every act can be performed in line with an individual's best knowledge at the time of the decision. The nexus is whether the best interest of

others is one's primary motivation. Analyzing one's own decisions based on the criteria of the other's best interest (cognitive/neocortex) eventually creates a new habit (instinctive/limbic brain). Eventually the buried instinct of altruism remerges in individuals, selfless action becomes natural.

7. Because brain structures involved in positive emotions, like compassion, are subject to change brought about by environmental input. A study of altruism found that children who have compassionate parents tend to be more altruistic. See Oliner and Oliner (1988). Children securely attached to their parents tend to be sympathetic to their peers, according to Waters et al. (1979). In contrast, George and Main (1979) note that abusive parents who resort to physical violence have less empathetic children and those whose parents use inductive corrections that seek understanding rather than punishment are more emphatic, according to Hoffman (1984).

8. In Theravada Buddhism, compassion (including self-compassion) is a power for deep mental purification, protection, and healing that supports inner freedom. In Mahayana Buddhism, compassion becomes the primary means to empower and communicate a non-conceptual wisdom in which self and others are sensed as undivided. In Vajrayana Buddhism, unconditional compassion radiates forth all-inclusively as a spontaneous expression of the mind's deepest unconditioned nature. Refer to Makransky et al. (2012).

9. For centuries, there have been attempts by philosophers to contradict, limit, expand, or modify it. Good summaries and discussions are found in Wattles (1996) and Cortesi (2001). Wattles' summary is pertinent and probably the way forward: "For responding to all these objections, there are three possible strategies: abandon the rule, reformulate it, or retain it as commonly worded, while taking advantage of objections to clarify its proper interpretation. I take the third way" (Wattles 2013).*

10. Each of them is also linked to the four stages of personal growth described along the book (awareness, acceptance, alignment, and accountability). These are further explained in Sect. 1.4.

11. More specifically, cognitive dissonance occurs when a person holds contradictory beliefs, ideas, or values and participates in an action that goes against one of these three. The experience of that expression causes psychological stress. When either of the following pairs—action/action, action/ideal, ideal/ideal—is psychologically inconsistent we do subconsciously all we can to either exit the situation that causes the clash, or by changing either part of the pair to establish coherence. The discomfort is triggered by the clash of the existing beliefs/values with new information (Festinger 1962).

12. An interesting connection with the foundation of our institutionalized religions can be established when we look at the origins of the semantics

related to 'beliefs.' In the English language, the word "belief" originally meant 'to love, to prize, to hold dear'. In the twelfth century, it narrowed its focus to 'the intellectual assent to a set of propositions, a credo'. It is useful to remember that "I believe" did not mean, "I accept certain creedal articles of faith." but "I commit myself. I engage myself" (Armstrong 2008).

13. The Golden Rule may come in various formulations: (i) Treat others as you would like others to treat you (positive or directive form); (ii) do not treat others in ways that you would not like to be treated (negative or prohibitive form); (iii) what you wish upon others, you wish upon yourself (empathic or responsive form).

14. Mental wellbeing is not narrowly located in the head but is assimilated by the physical body and intermingled with the natural world' (Hudson et al. 2019). Through the brain-gut-skin axis physiological factors like food or temperature influence our emotions and thus our thoughts—both directly (Holland et al. 1985) and indirectly via the microbiome that evolves in the gut (Taylor 2019).

15. I may be passionate about music, but unless I move beyond listening to practicing it myself, the pursuit of that passion does not entail that I have found and fulfilled my purpose. Conversely, I may feel that my purpose in life is to take care of my parents, and though I pursue this with diligence I am not passionate about it.

16. International trade, ODA, FDI, remittances, etc., belong to supra-national (interlinked) systems. For instance, international trade is the sum of bilateral exchanges which make up a whole system. However, it should be emphasized that the "agents" of these exchanges are (and their actions take place) in a country. This is so whether we refer to imports or exports, sending remittances, etc. Thus, when referring to this behavior, we locate it at the country (macro) level. As it is the case when individuals (micro) participate in a community (meso), their behavior is still at the micro-level.

17. While the meso-dimension is sub-national, certain institutions at the meso-level have counterparts at supra-national level (i.e., churches, multinational companies). This double nature illustrates the inseparable nature of dimensions which are connected through an ongoing dynamic of mutual interaction. A similar double-sized nature marks the meta-dimension which entails on the one hand supra-national institutions such as the United Nations, and on the other hand the sum of all entities, and spheres that are not human-related such as Nature.

18. These 4 components are in line with Cialdini's key principles of influence: reciprocity, commitment and consistency, social proof, authority, liking, scarcity (Cialdini 2006), and the subsequently added unity principle (Cialdini 2016).

19. Emotional inertia refers to the degree to which emotional states are resistant to change. Psychological maladjustment has been associated with both emotional under-reactivity and ineffective emotion regulation skills. The concept entails that overall emotion dynamics are characterized by high levels of inertia (Kuppens et al. 2010).

References

Abe, J. A. A. (2016).A longitudinal follow-up study of happiness and meaning-making. *The Journal of Positive Psychology, 11*(5), 489–498. https://doi.org/10.1080/17439760.2015.1117129.

Ambady, N., & Gray, H. M. (2002). On being sad and mistaken: Mood effects on the accuracy of thin-slice judgments. *Journal of Personality and Social Psychology, 83*(4), 947–961.

Ariely, D. (2008). *Predictably irrational.* Harper Collins. ISBN 978-0-06-135323-9.

Armstrong, K. (2008). My wish: The charter for compassion. TED. https://www.ted.com/talks/karen_armstrong_my_wish_the_charter_for_compassion.

Barrett, L. F. (2006). Solving the emotion paradox: Categorization and the experience of emotion. *Personality and Social Psychology Review, 10*(1), 20–46.

Baumeister, R. F., Vohs, K. D., Aaker, J. L., & Garbinsky, E. N. (2013). Some key differences between a happy life and a meaningful life. *The Journal of Positive Psychology, 8*(6), 505–516. https://doi.org/10.1080/17439760.2013.830764.

Baumeister, R. (1992, September 25). *Meanings of life* (Rev. ed.). New York: The Guilford Press.

Bechara, A., Damasio, H., Damasio, A. R., & Lee, G. P. (1999). Different contributions of the human amygdala and ventromedial prefrontal cortex to decision-making. *Journal of Neuroscience, 19*(13), 5473–5481.

Blackburn, S. (2001). *Ethics: A very short introduction* (p. 101). Oxford: Oxford University Press. ISBN 978-0-19-280442-6.

Bless, H., Fiedler, K., & Strack, F. (2004). *Social cognition: How individuals construct social reality.* Hove, East Sussex, UK: Psychology Press.

Borba, M. (2016). *UnSelfie: Why empathetic kids succeed in our all-about-me world.* New York: Simon & Schuster.

Cabanac, M. (2002). What is emotion? *Behavioural Processes, 60*(2), 69–83.

Carroll, L. (1865). *Alice's Adventures in Wonderland.*

Chardin de, P. T. (1955). *The phenomenon of man.* Harper Perennial Modern Thought (2008).

Cialdini, R. B. (2006). *Influence: The psychology of persuasion.* Harper Business.

Cialdini, R. B. (2016). *Pre-suasion: A revolutionary way to influence and persuade.* New York: Simon & Schuster. ISBN 978-1501109799.

Cohen, M. (1994). *Counseling and nature: A greening of psychotherapy.* Retrieved November 2020. https://files.eric.ed.gov/fulltext/ED374387.pdf.

Cortesi, D. E. (2001). *Secular wholeness.* Trafford on demand publishing.

Damasio, A. R. (2012). *Self comes to mind: Constructing the conscious brain.* New York: Vintage.

de Waal, F. (2005). *The evolution of empathy.* Retrieved September 2020 from https://rb.gy/ab4yww.

Diggs, B. J. (1973). The common good as reason for political action. *Ethics, 83*(4), 283–284. https://doi.org/10.1086/291887.

Doidge, N. (2007). *The brain that changes itself: Stories of personal triumph from the frontiers of brain science.* New York, NY: Viking Press.

Dewey, J. (1897, January). My pedagogic creed. Number IX in *The School Journal, LIV*(3), 77–80.

Ekman, P., & Friesen, W. V. (1971). Constants across cultures in the face and emotion. *Journal of Personality and Social Psychology, 17*(2), 124–129.

Ellis, L. (2016). *Engage with honor: Building a culture of courageous accountability.* Freedomstar Media.

Festinger, L. (1962). Cognitive dissonance. *Scientific American, 207*(4), 93–107.

Flew, A. (Ed.). (1979). *The golden rule. A dictionary of philosophy* (p. 134). London: Pan Books in association with The Macmillan Press.

Frankl, V. E. (1985). *Man's search for meaning.* New York: Simon & Schuster.

Frijda, N. (2004). Emotions and action. *Journal of Organic Chemistry.*

George, C., & Main, M. (1979). Social interactions of young abused children: Approach, avoidance, and aggression. *Child Development, 50*(2), 306–318.

Gino, F. Kouchaki, M., & Galinsky, A. D. (2015). The moral virtue of authenticity: How inauthenticity produces feelings of immorality and impurity. *Psychological Science, 26*(2015), 983–996.

Goetz, J. L., Keltner, D., & Simon-Thomas, E. (2010). Compassion: An evolutionary analysis and empirical review. *Psychological Bulletin, 136*(3), 351–374. https://doi.org/10.1037/a0018807.

Goldman, B. M., & Kernis, M. H. (2002). The role of authenticity in healthy psychological functioning and subjective well-being. *Annals of the American Psychotherapy Association, 5*(2002), 18–20.

Goleman, D. (1998). *Working with emotional intelligence.* New York: Bantam Books.

Gould, S. J. (1989). *Wonderful life.* New York: W. W. Norton & Co.

Graham, B. (1998). *On technology and faith.* TED. https://www.ted.com/talks/billy_graham_on_technology_and_faith.

Haidt, J. (2006). *The moral roots of liberals and conservatives.* TED. Retrieved August 2020. https://rb.gy/fep1gt.

Hoffman, M. L. (1984). Interaction of affect and cognition in empathy. *Emotions, Cognition, and Behavior*, 103–131.

Holland, R. L., Sayers, J. A., Keatinge, W. R., Davis, H. M., & Peswani, R. (1985). Effects of raised body temperature on reasoning, memory, and mood. *Journal of Applied Physiology, 59*(6), 1823.

Hudson N. W., Lucas, R. E., & Donnellan, M. B. (2019). Healthier and Happier? A 3-Year. Longitudinal Investigation of the prospective associations and concurrent changes in health and experiential well-being. *Personality and Social Psychology Bulletin, 45*(12), 1635–1650. https://doi.org/10.1177/0146167219838547.

Hume, D. (1740). *A treatise of human nature* (Oxford Philosophical Texts) (D. Fate Norton & M. J. Norton, Eds.). Oxford: Clarendon Press (2000).

Ibarra, H. (2015). *Act like a leader, think like a leader.* Harvard Business Review Press.

Johnson, S. (2019). *How 'extinction neurons' help us block out our worst memories.* Retrieved April 2020. https://bigthink.com/surprising-science/ptsd-memory.

Kahneman, D. (2011). *Thinking, fast and slow.* London: Penguin Books.

Kingsley, D. (2011). *Is technology rewiring our soul?* https://www.huffpost.com/entry/technology-spirituality_b_854757.

Korb, A. (2015). *The upward spiral: Using neuroscience to reverse the course of depression, one small change at a time.* New Harbinger Publications.

Kragel, P. A., & LaBar, K. S. (2016). Decoding the nature of emotion in the brain. *Trends in Cognitive Sciences, 20*(6), 444–455. https://doi.org/10.1016/j.tics.2016.03.011.

Kuppens, P., Allen, N. B., & Sheeber, L. B. (2010). Emotional inertia and psychological maladjustment. *Psychological Science, 21*(7), 984–991. https://doi.org/10.1177/0956797610372634.

Lehrer, J. (2010). *How we decide.* Boston: Mariner Books and Houghton Mifflin Harcourt.

Makransky, J., Germer, C. K., & Siegel, R. D. (2012). Compassion in Buddhist psychology. In *Wisdom and compassion in psychotherapy: Deepening mindfulness in clinical practice* (pp. 61–74). New York, NY: Guilford Press.

Marcus, G. F. (2005). What developmental biology can tell us about innateness. In P. Carruthers, S. Laurence, & S. Stich (Eds.), *The innate mind: Structure and content.* New York: Oxford University Press.

Marsh, J. (2009). *The unselfish gene?* Retrieved November 2019. https://greatergood.berkeley.edu/article/item/the_unselfish_gene.

Marsh, J. (2012). *Do mirror neurons give us empathy?* Retrieved March 2019 from https://greatergood.berkeley.edu/article/item/do_mirror_neurons_give_empathy.

Mullin, G. H. (1983). *Selected works of the Dalai Lama III. Essence of refined gold* (2nd ed., 1985). New York: Snow Lion Publications, Inc. ISBN 0-937938-29-7.

Newkirk, A. (2014). *The interactions of heuristics and biases in the making of decisions*. Retrieved October 2020. https://rb.gy/vs7ldp.

Nowak, M., & Highfield, R. (2011). *Supercooperators: Altruism, evolution, and why we need each other to succeed*. New York: Simon & Schuster.

Oliner, S. P., & Oliner, P. M. (1988). *The altruistic personality: Rescuers of Jews in Nazi Europe*. New York: The Free Press.

Oosterwijk, S., Lindquist, K. A., Anderson, E., Dautoff, R., Moriguchi, Y., & Barrett, L. F. (2012). States of mind: Emotions, body feelings, and thoughts share distributed neural networks. *NeuroImage, 62*(3), 2110–2128.

Parliament of the World's Religions—Towards a Global Ethic. (1993). Link to (PDF) via. www.parliamentofreligions.org/_includes/FCKcontent/File/TowardsAGlobalEthic.pdf.

Peters, A., Mc Ewen, B. S., & Friston, K. (2014). *Uncertainty and stress: Why it causes diseases and how it is mastered by the brain*. https://doi.org/10.1016/j.pneurobio.2017.05.004.

Rock, D. (2020). *Your brain at work: Strategies for overcoming distraction, regaining focus, and working smarter all day long* (Rev. and Updated ed.). Harper Business.

Ross, W. D., & Brown, L. (2009). *The Nicomachean ethics*. Oxford and New York: Oxford University Press.

Seligman, M. E. P. (1991). *Learned optimism: How to change your mind and your life*. New York, NY: Pocket Books.

Seligman, M. E. P. (2011). *Flourish: A visionary new understanding of happiness and well-being*. Free Press.

Seppala, E. (2013). The compassionate mind: Science shows why it's health and how it spreads. *APS Observer, 26*(5).

Scherer, K. R., Schorr, A., & Johnstone, T. (Eds.). (2001). *Appraisal processes in emotion: Theory, methods, research*. Oxford, UK: Oxford University Press.

Simms, K. (2011). The concepts of common good and public interest: From plato to biobanking. *Cambridge Quarterly of Healthcare Ethics, 20*(4), 554–562. https://doi.org/10.1017/s0963180111000296.

Stallone, A. (2018). *The most common cognitive biases*. Retrieved September 2019 from https://medium.com/@Angy_Stallone/the-most-common-cognitive-biases-9499bbbf97c7.

Stead, W. E., & Garner Stead, J. (1994). Can humankind change the economic myth? Paradigm shifts necessary for ecologically sustainable business. *Journal of Organizational Change Management, 7*(4), 15–31.

Strauss, C., Taylor, B. L., Gu, J., Kuyken, W., Baer, R., Jones, F., & Cavanagh, K. (2016). What is compassion and how can we measure it? A review of definitions and measures. *Clinical Psychology Review, 47*, 15–27.

Strohminger, N., Knobe, J., & Newman, G. (2017). The true self: a psychological concept distinct from the self. *Perspectives on Psychological Science, 12*(2017), 551–560.

Sunstein, C. R. (2002). Thinking about risks. In *Risk and reason: Safety, law, and the environment* (pp. 28–58). Cambridge, UK: Cambridge University Press.

Taleb, N. N. (2007). *The black swan: The impact of the highly improbable.* Random House. ISBN 978-1400063512.

Taylor, V. H. (2019). The microbiome and mental health: Hope or hype? *Journal of Psychiatry & Neuroscience: JPN, 44*(4), 219–222. https://doi.org/10.1503/jpn.190110.

Thagard, P., & Aubie, B. (2008). Emotional consciousness: A neural model of how cognitive appraisal and somatic perception interact to produce qualitative experience. *Consciousness and Cognition, 17*(3), 811–834.

Thaler, R. H., & Sunstein, C. R. (2008). *Nudge: Improving decisions about health, wealth, and happiness.* New Haven, CT: Yale University Press.

Tversky, A., & Kahneman, D. (1982). Judgment under uncertainty: Heuristics and biases. In D. Kahneman, P. Slovic, & A. Tversky (Eds.), *Judgment under uncertainty: Heuristics and biases* (pp 3–20). Cambridge, UK: Cambridge University Press.

Walther, C. (2020a). *Development, humanitarian action and social welfare.* Macmillan Palgrave. New York.

Walther, C. (2020b). *Humanitarian work, social change and human behavior.* Macmillan Palgrave. New York.

Waters, E., Wippman, J., & Sroufe, L. A. (1979). Attachment, positive affect, and competence in the peer group: Two studies in construct validation. *Child Development*, 821–829.

Watson, L., & Spence, M. T. (2007). Causes and consequences of emotions on consumer behaviour: A review and integrative cognitive appraisal theory. *European Journal of Marketing, 41*(5/6), 487–511.

Wattles, A. (1996). *The golden rule.* Oxford University Press.

Wattles, A. (2013). The golden rule. Don't take it literally. *Psychology today.* Retrieved September 2020. https://rb.gy/g5ft8f.

Winerman, L. (2005). The mind's mirror. *Monitor on Psychology, 36*(9). The American Psychological Association. http://www.apa.org/monitor/oct05/mirror.

CHAPTER 2

Optimization

Abstract Every human being is a micro-representation of the continuum of tangible and intangible factors that influence Society. *Acceptance* of the dimensions that shape existence, with or without technology, is the second of the 4 stages (awareness, acceptance, alignment, accountability) that the mindset needs to traverse for personal change and through it, large-scale social transformation. This chapter makes the case that technology can serve the existing interplays to bring out the best in and for individuals, and thus for society. We analyze why and how technology can be harnessed to not only influence but optimize the multidimensional dynamic, if the underpinning human mindset is geared toward the common good. Optimizing who we are, individually and collectively, does not depend on technology; but our ability to optimize technology to serve us in this endeavor is conditioned by our state of mind without artificial tools. Technology can be beneficial to humankind only if the natural human ability to feel, think, and aspire has been acquired beforehand.

Keywords Optimization · Coherence · Acceptance · Values · Power · Technology

The tower of knowledge becomes higher during each generation, which expands the view to be had from above and increases the risk of collapsing.

© The Author(s), under exclusive license to Springer Nature Switzerland AG 2021
C. C. Walther, *Technology, Social Change and Human Behavior*,
https://doi.org/10.1007/978-3-030-70002-7_2

41

From the collapse of ancient Rome to the fall of the Mayan empire, archaeological evidence suggests that five factors were almost invariably involved in the loss of civilizations: uncontrollable population movements; new epidemic diseases; failing states leading to increased warfare (or internal strife); collapse of trade routes; and climate change. In one form or another, 'technology' was involved as a factor that fastened decline (Hutt 2016). Every generation felt part of the cusp of a new era—perceiving the stage of technological prowess as the peak ever to be reached. Like we do today.

Technology refers to the development and use of basic tools to improve human experience.[1] It can be described as 'the sum of techniques, skills, methods, and processes used in the production of goods or services or in the accomplishment of objectives, such as scientific investigation' (Bain 1937). On the one hand, it relates to the knowledge of techniques and processes; on the other, it may be embedded in machines and systems to allow for operations to function without detailed knowledge of their workings[2] (Hughes 2004). Simply said technology is (1) a means to fulfill a human purpose, (2) an assemblage of practices and components, and (3) a collection of devices and engineering practices available to a culture (Arthur 2009). In the current context, it is important to remember the first point—it is 'a mean' to fulfill a (human) end. The trajectory toward a future in which humans serve robots is an oxymoron which illustrates a purpose that was defeated by itself—possibly for lack of awareness of the workings behind the scenes.

For many, Technology is a 'black box.' Used variably by economists, engineers, and psychologists, the "black box" expression reflects how technological phenomena are commonly treated as events that transpire from the insides of an unknown, circumscribed space (a black box), and "are addressed under the self-imposed ordinance not to inquire too seriously into what transpires inside that box" (Rosenberg 2008). Here we peer not only into the black box of technology but also into the black box of human experiences and expressions. As we will see both may be treasure chest or Pandora's box.

The more sophisticated a tool is—be it an algorithm or a piece of equipment—the more important it is that the underpinning vision for its use is clearly set. To be 'clearly set' in the context of the POZE paradigm means two things—it is coherent in terms of its functionality and its goals, but more importantly, it is squarely anchored in the ambition of a positive impact on humanity, in the short, medium, and long term. Humanity and technology are connected within an ongoing feedback-loop. Technology can help in shaping a society where people are lifted to fulfill their potential, if those who constitute that Society (Us) aspire to that objective in the first place—and act accordingly.

Technology has influenced our thoughts and behavior throughout the course of history. Conversely, the shape of technologies that we are seeing today was influenced by the thoughts and desires that prevailed at the time when they were made. Technology changes what we do, where we are, and how we interact. The prehistoric discovery of how to control fire and the subsequent Neolithic Revolution increased the availability and preservation of food, and the invention of the wheel helped humans to travel and transport materials in ways that exponentially expanded what the human body could accomplish. The telegraph, the telephone, and the internet have lessened physical barriers to communication and allowed humans to interact on a global scale instantaneously. Technology influences WHO we are and become.

Throughout history, technology has opened additional venues of influence that resulted in new patterns within the human psyche. Each gain had costs. Examples of this process include the introduction of cuneiform alphabets and tablets among the earliest civilizations and the Gutenberg printing press; they made information more available within the public domain (even though at the time most people were illiterate). This in turn affected the psychological and physiological conditions of those exposed to printed information. Not only did the general masses begin to read on a large scale but the very act of reading stimulated parts of the human brain which had been used differently before. The increase of a reading public and the expansion of schooling decentralized on the one hand access to knowledge, which contributed to tendencies of individualism and instances of opposition to ruling structures; on the other hand, education has been instrumentalized ever since to inculcate and tame entire generations from an early age (Kingsley 2011). Today some argue that the internet is not only shaping our way of living, but how we think and behave by physically altering our brains (Carr 2008).

Technology is one asset among others to maximize the outcome of the *mmmm-matrix*. It can be a game-changer, but it does not replace the foundations that make us who we are.

Technology enables us to gather and scrutinize immense sets of data to detect patterns. Deriving from the investigation of gigantic data quantities, conclusions and recommendations may be drawn to optimize the interplay that influences our individual and collective existence. However, we must accept two limitations of our expanding knowledge:

1. Despite ever larger scientific quantum leaps, much remains beyond the grasp of human knowledge. Knowing what we do not know may be our biggest asset when it comes to understanding. "To err is quantum, to correct is divine" (Cho 2020).
2. Never can technology exempt individuals from their personal accountability to act as aspired in favor of the common good; otherwise, it turns from an open door of opportunity into a prison of mind and matter.

The following pages refer to technology, both software and hardware, as a tool that involves energy, but more precisely information. From phone to internet, from artificial intelligence (AI) to smart devices, we look at the status quo to argue for the human factor as the centerpiece of humane technology.

2.1 Technology: Treasure Chest—Black Box—Pandora's Box

To optimize technology, we must be aware of its influence on us and others. This is facilitated when we understand the intentions that underpin the design of our tools. Whether technology benefits humanity or puts it in danger depends ultimately on those who design it, and those who use it. This section explores both options.

Analyzing data from the past 120 years, certain researchers now argue that humanity has not only reached its peak but is likely to decline. There appear to be limits on characteristics such as how tall, strong, and fast a human being can be. Furthermore, our impact on the environment, including pollution and global warming, is likely to pull those limits further down. "This will be one of the biggest challenges of this century as the added pressure from anthropogenic activities will be responsible for damaging effects on human health and the environment" (Toussaint in Griffin 2017).

The inherent challenge of these and other projections for the future is that they are based on the past. Such a perspective is shortsighted, since the unknown cannot be envisioned based on the acquired. "When you develop your opinions on the basis of weak evidence, you will have difficulty interpreting subsequent information that contradicts these opinions, even if this new information is obviously more accurate" (Taleb 2007). The same holds true about the future of humanity and our respective lives. When it comes to future outcomes, one step taken today with a fresh mindset can be more meaningful than ten steps taken yesterday. Purpose matters to guide steps that matter for our individual and collective wellbeing.

2.1.1   What Is the Role of Technology in the Quest for Wellbeing?

Hardware is required to run software, while software is needed to operate hardware. Human wellbeing involves material and immaterial dimensions, which mutually influence each other. The range of needs is reflected in major human rights conventions such as the Universal Declaration of Human Rights, the Convention of the Rights of the Child, or more recently the Millennium Development Goals and their successor the Sustainable Development Goals. All of them embrace material aspects and immaterial elements that matter for the Quality of one's life (Tonon 2015; Michalos 2014). Yet these catalogues are not complete nor all encompassing. They lack the entitlement to live in a society that lifts individuals to fulfill their inherent potential. For the remainder of this book, we will consider the range of human rights as an acquired commitment made by States.[3] They are the basic minimum of human development. Rather than dwelling on them, the aim here is to go beyond, to an active obligation for those who hold (in)formal power to use the latter for those who lack it, and furthermore, to systematically channel their means to cultivate the common good (Walther 2014).

Access to technology increases rapidly throughout the world. However, this does neither imply that it is distributed equally, nor that the results of such access benefit everyone in the same way.

The uses and outcomes of technology illustrate two major dynamics that characterize society—the disconnect of young and old, and the gulf that separates rich and poor. On the one hand, Generation Z in high-income countries is now of age to work and vote. In difference to the generations before them, they never experienced a world where smartphones, computers, online shopping, and wearables were not omnipresent. Generational cutoff points are neither an exact science, nor are their boundaries completely arbitrary (Dimock 2019). Technology, in particular the rapid evolution of how people interact, is a generation-shaping consideration. Even though Gen Z cohabitates with Millennials (born between 1981 and 1996), Generation X (born between 1965 and 1980), Baby Boomers (born between 1946 and 1964), and those born before 1946, their perception of 'reality' has been shaped from an early age through the filter of technology.[4]

The resulting influence on youth behavior, attitudes, and lifestyles is both positive and concerning (Dimock 2019). Overall, digital technology use has stronger effects on short-term markers of hedonic wellbeing (i.e., negative affect) than long-term measures of eudaimonic wellbeing (i.e., life satisfaction) (Dienlin and Johannes 2020). Although adolescents are more vulnerable, effects are comparable for both adolescents and adults. It appears that both, low use and excessive use are related to decreased wellbeing, whereas moderate use is related to increased wellbeing. These results are aligned with the so-called set-point hypothesis, which posits that life satisfaction varies around a fixed level, showing much interpersonal but little intrapersonal variance through time (Lucas 2007).

In the absence of long-term studies, it remains to be seen whether these are lasting generational imprints or characteristics of adolescence that will become more muted over the course of adulthood. In the meanwhile, the aspirations and actions of Gen Z influence the course of society. They evolve in parallel to older individuals who often struggle to cope with the increasing necessity of familiarizing themselves with technology. In an always-on culture, those who are off the grid are off the charts. The classic principle 'out of sight, out of mind' applies—those who are not able or willing to join the online circus are at risk of falling behind. This is even more acute for people living in locations that remain barely touched by technology

As of July 2020, almost 4.57 billion people were active internet users, encompassing 59% of the global population. Mobile has now become the most important channel for internet access worldwide as mobile internet users account for 91% of total internet users (Clement 2020).[5] COVID-19 accelerated the shift to digital forms of interaction, fundamentally shaking up how businesses and individuals operate (Fitzpatrick et al. 2020). During the pandemic, the internet has become essential to everyday life in many places due to lockdowns and a massive shift from office work to work-from-home modalities (Oluwaseun 2020). But even in so-called high-income countries such as the United States, millions of people do not benefit from the recent technological advances for lack of stable access to the internet or due to the inability of affording the necessary hardware—or because they lack the needed skills.

The technological no-mans-land is shrinking every year; yet as of July 2020, 41%, nearly 1 in 2 people had no internet whatsoever (Clement 2020). In many low-income countries, this deprivation adds to a long list of unaddressed fundamental needs. In times of waste and abundance, billions remain deprived of food, running water, electricity, health care, education, and social security. The world is evolving at a paradoxical speed; one part is still struggling to cope with problems that have haunted humankind for centuries, whereas the other part seeks to cope with the impact of problems that were supposed to be solutions. To help both parts move beyond the present conundrum, where one step forward eventually represents one step backward, technology must be designed and used in a holistic understanding.

2.1.2   *From Blessing to Burden*

While millions are deprived of it, for others technology has become a source of distress. Constant usage and with it the unescapable opportunity of 'connection' has turned into a curse. We have become voluntary slaves of the tools that were created to serve us.[6] Constantly navigating the stream of news from 'friends' on social media, including a large proportion whom they never met, especially young people feel constant pressure to live up to the same exhilarating experiences that these friends share. Reading what others are doing while being a bore at home feels unbearable for many (Kavi et al. 2013).

Two decades ago, bullying was limited to face to face and thus a relatively short time window; now that space has grown exponentially, in terms of time, space, and torture tools.[7] The targeted individual has no place to hide on the grid; and while in the past words and drawings were enough, now multimedia is available with a panoply of platforms to share the exploits. The past left it possible for victims to destroy the trail of pain; the internet's memory is illimited.

Why do our inhibitions fade in the virtual space? What one would never tell even close relatives or friends becomes a topic of conversation with strangers in a chat room. Rather than enhancing or complementing offline connections, the latter are increasingly replaced by virtual experiences. In the attempt of feeling alive in a parallel world, the real thing has begun to fade; which makes the online space ever more appealing. We have ever more opportunities to spend our time, yet we are running out of it. Like money, lifetime is limited. 24 hours are wired to our account every morning. How much of it do you spend online? Time-management apps, especially those geared to restrict our online time, have grown into a gigantic market. Google has a whole battery of people to come up with innovations that help people break the dependency from their devices (one is tempted to wonder whether these come with the same keen success intention as cigarette publicity that emphasizes the danger of smoking). It is a worrying landscape, but there is more. The influence of technology evolves to be ever more subtle and intense.

2.2   Artificial Awareness

Current intelligent systems can treat tremendous amounts of data and make complex calculations fast. Yet they lack an element that is key to building sentient machines—Awareness. The full extent of AI will come to play when self-aware AI starts to build self-aware AI, skipping the human maker in the process to further sophistication (Hintze 2016). To reach that stage, however, human engineers themselves must understand human consciousness better.

There are 4 different types of artificial intelligence: reactive machines, machines with limited memory, machines with an in-built theory of mind, and machines imbued with self-awareness. The first two exist already, and the other two are in the making.

Reactive AI does not interactively participate in the world. For instance, although it improved its future moves based on the result of past ones, Deep Blue, IBM's chess-playing supercomputer, which beat international grandmaster Garry Kasparov in the late 1990s, falls in this category.[8] The computer can 'see' the world and act on what it sees, without an internal notion of that world. Google's AlphaGo has beaten top human Go experts with a capacity that is several magnitudes beyond the human brain. Using a brain-like neural network to evaluate game developments, it is far more sophisticated than Deep Blue, yet it still cannot evaluate all potential future moves. For lack of a concept of the wider world, these machines cannot function beyond the tasks that were assigned to them. They can accomplish a mission but lack a vision.

In a way, this type resembles a human who is following the trend of inertia. Triggered into repeating the same habit in the present by drawing on experience from the past, it applies outdated thought patterns to new situations.

Limited memory AI can investigate the past to interact (or help us interact) with the world. For example, self-driving cars observe the speed and direction of other cars, identifying specific objects and monitoring them over time. These observations are added to the cars' preprogrammed representations of the world, which include lane markings, traffic lights, and other traffic-relevant elements, like curves in the road. These aspects are activated when the car decides when to change lanes to avoid crashing into another driver. However, these pieces of information are transient. Unlike human drivers who compile experience from years behind the wheel, these pieces are not saved as part of the car's library of experience to learn from over time.

Building AI systems that can draw on a full representation of the world,[9] remember their experiences, and learn from them to handle new situations is challenging. So far, the result is restrained by human limitations. Potentially, AI refinement may benefit from the theory of Darwinian evolution; the trick is to make up for human shortcomings by building machines that can build their own representations (Marstaller et al. 2013). However, even when this will be accomplished, AI will still be missing a theory of the mind.

AI that is gifted with theories of mind refers to machines that not only form representations about the world, but also about other agents and entities in that world. The psychological concept of theory of mind entails the awareness that people can have thoughts and emotions that affect their own behavior. This understanding is essential to the formation of human societies; it is at the core of social interactions. Without the ability to (at least attempt to) understand other people's intentions, and without considering the perspective of others about oneself and the shared environment, collaboration is hard if not impossible.

Many individuals not only fail to understand their fellow human beings but lack the willingness to feel with them (empathy) and based on these feelings, to do something for them (compassion). The prevailing global cacophony of greed and gaps illustrates that.

Self-aware AI development entails the conceptualization and creation of systems that form representations about themselves. For achieving this, AI engineers must not only understand consciousness but build machines that have it. Conscious beings are aware of themselves, know about their internal states, and can (attempt to) predict the feelings of others. When someone is crying, we assume that they are sad, because that is how we usually feel when we cry (Mirror neurons are at play. See Chapter 1). Without a theory of mind, we could not make these sorts of inferences. We learn about ourselves and others based on our own experiences. By analyzing our intentions, thoughts, and sensorial impressions, through the POZE prism, connections emerge. Gradually, our sensitivity to the prevailing interplays sharpens, and with it our ability to optimize them consciously.

Self-aware machines are not at reach, yet. The questions that must be addressed in the meantime are relevant not only for CEOs, researchers, and software designers. The refinement of our PERSPECTIVE of human minds and motivations determines the design and orientation of artificial intelligence. Who we are and why we create technology ultimately influences what technology does, and thus where the machines that we create will take us.

2.3   Algorithmic Influence

Technology increasingly impacts all aspects of human life: our mind and bodies, our emotions, and our aspirations. Paradoxically, we are, on the one hand, acutely aware of the all-pervasiveness nature that technology has developed, and, on the other hand, we are (willfully) blind to the depth and extent of algorithmic influences on our beliefs, opinions, and choices.

Algorithms are aimed at optimizing everything. They are instructions for solving a problem or completing a task (Rainee and Anderson 2017). Everything that we see and do on the web is a product of algorithms. Whatever ends up in a person's social media feed is brought there by algorithms. From Computer codes to the Internet, via emailing and online searching down to Smartphone apps, GPS navigation, and computer games, algorithms are everywhere. Whether it is the proposition of a soul mate, a movie to watch, or a holiday destination, online platforms operate through algorithms. Artificial intelligence thus influences how we perceive the online world and our natural environment. Increasingly, algorithms are developed to help us make choices in our personal and professional lives: how, when and where we exercise, eat; and spend our money; who we interact with; etc. The acute risk is that AI is becoming ever stronger whereas our natural intelligence weakens for lack of exercise. The brain is a muscle that must be challenged to grow. It can heal from external damage (Doidge 2015) but it is disarmed when confronted with the harm that comes from lack of use.

Cognitive overload as a result of a 24/7 sensorial experienced combined with inertia results in the continuous discarding of information. We know that we do not retain most of the flurry of newsbytes that we take in every online moment, but that does not change our desire to keep on scrolling and browsing. Extensive research about the workings of our brain and behavior is available. Yet such access to information does in no way ensure that people use it (Risdon 2017). However, leveraging behavioral insights serves to fine-tune algorithms. In the best-case scenario, this may lead to "empowering individuals, in all their cognitive and behavioral diversity, to make better decisions based on their own data and preferences" (Nioeber and Welsh 2020). Yet instrumentalized with commercial or political interests, the combination of psychology, neuroscience (the in-depth understanding of the mechanical workings of the brain), and machine learning is a ticking time bomb.

2.3.1 Priming Power

Priming illustrates the subtle line between the benefits and burden of influence. It is a phenomenon whereby the presence of one stimulus influences a response to a subsequent stimulus, without conscious guidance or intention (Weingarten et al. 2016). For example, the word EMOTION is recognized more quickly following the word HEART than following the word APPLE. Priming can be perceptual, associative, repetitive, positive, negative, affective, semantic, or conceptual (Bargh and Chartrand 2000). Even subtle suggestions to the subconscious mind can influence subsequent behavior. In one experiment, some subjects read a list of words that included some words related to old age and infirmity (e.g., "gray" and "wrinkled"). These subjects were found to walk more slowly to the building's elevator after the apparent conclusion of the experiment than a control group that had not read the age-related terms (Gladwell 2007). Illustrating the fluid interplay of brain and body, internal and external influence, and their impact on human behavior, priming is an example of the POZE paradigm's principle of a mind-matter continuum that reflects decision-making as a process that goes far beyond the cognitive sphere.

Priming is different from presenting the target audience with information, appealing to their needs (real and perceived), and offering tangible and intangible benefits. Zooming in on the subconscious, the success of priming derives from activating the user's intricate mix of memories from the past, present emotions, and expectations for the future. That power to influence subconscious decision-making processes has gigantic potential for (ab-)use of all sorts. Already, it is used widely by Amazon & Co to trigger the desire to consume their products. Similarly, social media platforms filter which posts, articles, and advertisements they show in your feed, in line with your previous reading choices. Past interests thus lay the ground for future access to information and thereby influence the opinions that derive from this information.

Environment matters when it comes to behavior (Kahneman 2011). Not only the presence and absence of other people and their behavior influence our experiences and expressions (a fact that is increasingly understood and used through the help of algorithms (Chen et al. 2011), but also, purely sensorial, verbal, or circumstantial factors matter. Thus, often the most effective approach to influence a person is a combination of multi-channel messages (classic communication) and choice architecture tweaks (behavioral economics approach[10]). As algorithms become ever more sophisticated and pervasive, individuals are increasingly inundated with digital nudges (Perrott 2020). Machine learning makes it possible to tailor marketing approaches specifically to the individuals that are part of target audiences.[11] Due to the all-pervasiveness of technology, we are primed to engage in certain behaviors without being consciously aware of the trigger that induced the preceding emotion.

Priming falls outside the *Scale of Influence* (Sect. 1.4) and at the same time covers all 4 stages. Even though the effectiveness of priming decreases if the subjects are aware of the cue's expected influence on them (Bargh and Chartrand 2000), it may still operate. Under certain conditions, we can even prime ourselves toward certain behaviors.[12] Priming techniques may thus serve to optimize the complementary impact of inside-out and outside-in dynamics (Sect. 1.4).

2.3.2   Facts Versus Opinions

In a Society that is technologically 'connected' 24/7, it is vital to distinguish between original opinions, second-hand opinions, and third-hand opinions. The first relates to those perspectives of ours that are genuine, resulting from the effort of personal thinking about an issue; the second refers to the pickup of opinions that others put out, bilaterally or in the public space; we adopt them as ours because they seem to resonate with our overall value system; the third type is the acceptance of forwarded opinions that others had picked-up and shared, without much deliberation. With each type, the level of conscious deliberation and the number of people who genuinely make up their own mind about the issue decrease. Such a decrease is not conducive to intellectual freedom, which is the core of democracy. "A popular government, without popular information, or the means of acquiring it, is but a prologue to a farce or a tragedy; or perhaps both. Knowledge will forever govern ignorance; and a people who mean to be their own governors, must arm themselves with the power which knowledge gives" (Madison in Ball 2003). With the abundance of opinions on the web, the risk of thought inflation increases and with it the normalization of adopted perspectives as a natural sequel to the homegrown effort of personal deliberation. "Where all think alike, no one thinks very much" (Lippmann 1915).

Information from social media is widely accessible, decentralized, and thus theoretically exposing us to a wide range of viewpoints, beliefs, and narratives. However, algorithm filtering, systematic manipulation of online debates, and the strategic dissemination of misinformation render this exposure a two-sided sword, with potentially deadly consequences. The year 2020 saw the vicious cycle of a health pandemic caused by the coronavirus fueled by an infodemic of rumors and false information. "The inability to cordon off the basic facts of public health from reactionary propaganda threatens people's lives" (Philipps 2020). The separation of facts and opinions, of beliefs and wishful thinking, is becoming opaque when uncertainty about the future is the name of the game. It is a fertile ground for demagoguery.

The online market-square is marked by "the tendency to dissolve complex policy issues in a blinding moral certainty" which makes "everyone less capable of democratic participation" (Harper's Letter 2020).[13] In addition to such shallow repacking, algorithms expose users primarily to opinions that echo their own. AI plays into our brain's tendency of confirmation bias. We are biologically attuned to believe what we believed before, especially when we are in a state of mental distress (Talluri et al. 2018).

Technology offers the potential of a global public square that is open to all opinions. But beyond rules and tools, a common value orientation such as the Golden Rule of reciprocity is required to sift good from evil (Wattles 2013). A purpose such as the common good may serve as a meta-filter to eliminate hate.

The link of conspiracy theories and individual wellbeing is a perfect illustration of the link between emotions and thoughts, but also between individual wellbeing and collective welfare. The COVID-19 pandemic fostered feelings of grief, uncertainty, powerlessness, and marginalization. It is a conducive environment for conspiracy narratives. Because unfulfilled psychological needs, including the need to understand the world, to feel safe, cared for, and good about oneself and one's social group, underlie conspiracy beliefs. Conspiracy theories entail a mix of thoughts about an issue, emotions that derive from it, and pre-existing believes. They undermine political participation, discourage environmental protection, and incite violence (Amarasingam and Argentino 2020). "Neglecting the mental-health crisis risks perpetuating an information one" (Cichocka 2020). Conversely, the spread of misinformation is masking healthy behaviors and promoting erroneous practices that increase the spread of the virus and ultimately result in poor physical and mental health outcomes among individuals. Myriads of incidents of mishaps caused by these rumors have been reported globally (Tasnim et al. 2020).

Algorithms can optimize or disrupt human interactions. For now, they are the cause and consequence of tremendous power and control in the hands of selected corporations and governments (Rainer and Anderson 2017). Our 'choices' equip them to perpetuate bias, create filter bubbles, eliminate alternative choices, and stifle serendipity. Unless we approach the online space with an open mind, consciously searching for opinions that contradict our own, we are prone to fall for 'facts' that suit our own perspective. The only way to break that spiral is to break out of the echo-chamber of our mental comfort-zone.

Algorithms recognize faces, sort photos, write text, build and drive cars, produce videos, and design game-plans. We are at the beginning. With the emergence of self-learning and self-programming algorithms, it is likely that future algorithms will write most algorithms themselves, with far-reaching consequences in the offline sphere.

The World is an interconnected Universe, covered by a mesh with ever smaller holes. Our attitude toward ourselves, society, and technology will influence whether we suffocate under this mesh. We can optimize it as a safety net to protect humanity from crashing into chaos. Changing the world, or even our own circumstances, is an intimidating challenge. But unpacking the conundrum at hand makes that task manageable.

The *mmmm-matrix* is not an intellectual endeavor. It is a pragmatic perspective to reverse-engineer change, starting with the status quo.

In the next section, we explore the unfulfilled potential that derives from technology. The theme of unfulfilled potential will then be pursued in Chapter 3, where we look at the unfulfilled potential of humans from the micro-, meso-, macro-, and meta-perspective. Awareness of the multi-dimensional repercussions of technology makes it easier to draw a line between ingenuity and infatuation with our skills, between creativity and complacency with the status quo, and between compassion for humanity and our own comfort-zone.

2.4   Technological Treasure Hunt

Technology can save and improve lives. Making it easier to connect, move and share regardless of physical boundaries such as language, distance, disability. It might serve to intensify our experiences and render us more aware of ourselves and the environment in which we evolve.

"If I have seen further it is by standing on the shoulders of Giants" (Newton 1675). The present results from the past and represents the stepping-stone of the future. The First Industrial Revolution used water and steam power to mechanize production. The Second introduced electric power, a range of new chemical innovations and processes, telegraph and telephones, and was accompanied by Taylorism and Fordism. The Third used electronics and information technology to automate production. Now a 4th Industrial Revolution (4iR) is happening. It is characterized by a fusion of technologies that is blurring the lines between the physical, digital, and biological spheres (Schwab 2019). With each wave, the human perception of time and space has evolved. Because when the time necessary to connect distinct geographical locations is reduced, distance is compressed, resulting in its gradual "annihilation" (Marx 1848). Our experience of space is intimately connected to the temporal structure of the means by which we experience space (Harvey 1989). "Space by itself, and time by itself, are doomed to fade away into mere shadows, and only a kind of union of the two will preserve an independent reality" (Minkowski 1906 in Corry 1997). Technology makes the relativity of reality apparent, while bringing to light the potential of change for universally enjoyed life quality.

Changing the world was never as naïve as it sounded. Even a brief overview of certain opportunities of the Fourth industrial revolution

(Marr 2019) illustrates what *could* be harnessed for positive social transformation. Experts predict that over the coming decades cost will be falling by at least 10× in key sectors including food, transportation, energy, materials, and information, as production processes become more efficient (Arbib and Seba 2020). These changes could help to eliminate poverty. The question is not whether it is possible to close the gap between needs and means, but if the willingness to do so exists. The question is not whether technological solutions related to the closing of the gap exist or will emerge. Rather, the point is that we must nurture the willingness to shepherd technology in that direction. Within a perspective of interconnectivity, a clear role emerges for every component of the *mmmm-matrix*. Individual attitude is the point of departure for change toward OPTIMIZATION.

The relationship between people and technology is not an either/or question. We must not decide if we want to move back to the caves or forward to the future. Answers are more complex. We must step out of shortsighted dynamics that seem to serve us but are detrimental in the long term, and move toward the conscious choice, design, and use of technology that is intended to optimize the *mmmm-matrix*.

Making the choice of inclusive progress is a straightforward decision once we accept that everything is connected. Because when we acknowledge that individuals, communities, countries, and the planet are part of one and the same universal equation, it is a logical (self-interested) choice to seek the OPTIMIZATION of all parts and their interplay. Limiting the scope to just one piece of the whole is like cutting your finger while looking away. Although the eyes do not see it, the body registers the pain acutely and the pain-receptors in the brain go in alarm-mode, resulting in frantic thoughts. Simultaneously, emotions of panic and fear (and possibly anger about your clumsiness) are triggered. The myopic scope that was pursued by many for many centuries has similar outcomes. By concentrating efforts on our own personal interests, we were hurting part of the organism that we are part of. While cutting a finger is clumsiness, disregarding the needs of others is a choice. Choices have consequences for which we are accountable.

A societal set-up where a few have (too) much, and many have (too) little, is erroneous, not only morally speaking as it goes straight against our innate moral foundations (Haidt 2006)—but in practical terms. Because in a society that is marked by inequality the overall life quality of everyone is reduced, including those who are on top of the social food-chain (Wilkinson and Pickett 2009).

To tailor and maintain a social setting that is conducive to everyone's life quality is a win-win-win-win. It is (i) morally right; (ii) in everyone's interest; (iii) sustainable and cost-efficient due to a lower environmental footprint; and it is (iv) rewarding—in terms of both the process itself and deriving from the outcomes. Relevant changes could happen without technology. Yet, designed and delivered with the right intentions, technology may serve to accelerate that process by connecting, expanding, and scaling thus far disconnected initiatives.

The following is a small selection of potentially beneficial technologies. They are regrouped in the 4 dimensions that we have seen earlier—body (hardware), mind, heart, and soul (software). However, as seen earlier, these 4 dimensions interact (Chapter 1). Consequently, the palpable and perceptible influences of technology intersect.

2.4.1   Body—Overcoming Physiological Limitations

Since 1900, the global life expectancy has more than doubled, reaching above 70 years on average. The inequality of life expectancy is still large across and within countries. In 2019, the country with the lowest life expectancy was the Central African Republic with 53 years, whereas in Japan people lived on average three decades longer (Roser et al. 2019). Better nutrition and hygiene together with medical progress are among the main causes of this progress. From soap over vaccination to cancer treatment; from identifying those at risk of a heart attack or diabetes to then reducing that risk with medicines and systematic life-style changes; from self-dissolving implants to heal a broken limb to artificial organs, technology has radically transformed the landscape of curative and preventative medicine. From bioengineering to designer babies, over at-home DNA mapping, and artificial organ farms, the lines that separate acquired, possible, and unfathomable technologies shift every month.

Health care is rebuilding services around the Internet of Things (IoT).[14] Medical check-ups involve ever more interactions with sensors, cameras, and robotic scanning devices and less with human doctors and nurses. Telemedicine makes it possible to carry out consultations from afar. Even treatment in far off locations which lack qualified doctors yet dispose of on-site medical staff is now possible thanks to tele-guidance based on remote imaging. Patient autonomy has increased due to (wearable) technology. Rather than depending primarily on their doctor for information and advice, people now increasingly refer to the Internet as a non-stop-shop. Often this makes patients feel more involved in assessing and addressing their health. Using app-based and wearable tools to monitor their health individuals can now self-diagnose a wide number of conditions at home. Personalized health metrics gathered through activity trackers and smartphones provide real-time information, which then triggers AI suggestions that may be conducive to prevent or manage certain conditions. Data from wearable devices increasingly offers the possibility to predict, detect, and treat health issues even before symptoms arise. Precision medicine refers to tailoring curative and preventive interventions that are in line with the specific health issues and needs of a person. It may be favored by our technological prowess, provided the deriving interventions are designed with a holistic mindset.

At the same time as we depend less on doctors, we hang on ever more to our devices. We know more about ourselves and understand it less. On the one hand, real-life data about the functions of our body can serve to educate the user. Because data gathered over time shows patterns. Your activity tracker shows that over a month time your heartbeat is over the normal resting rate in the evening, and that your sleeping time is underneath the required minimum. Putting pieces together you realize that a big expresso after dinner is not conducive to your health. The risk of data-tracking is that physiological features like heartbeat, food-intake, or steps walked become abstract metrics; something (else) to track without treasuring it; as we do with other 'things' in our life, from money to stamps. We come to rely on devices to track our body, rather than listening to its signals. By translating exercise, nutrition, and sleep into abstract metrical data, devices that could strengthen the relationship with our body are thus increasing the distance between our 'self' and our physiological shell. We focus on the outside and forget the internal counterpart.

2.4.1.1　The Line Between Body and Brain, Mind and Matter, Real and Imagined Becomes Blurry

Technology changes not only how but what we perceive via our senses, expanding and amending them ever more. So-called Extended Reality (XR) covers several immersive digital experiences, referring to virtual, augmented, and mixed reality: Virtual reality (VR) provides a digitally immersive experience where the visitor enters a computer-generated world using headsets that blend out the real world. Augmented reality (AR) projects digital elements (e.g., sounds or images or haptic [through touch] feedback) onto the real world, thus combining real and digital experiences. Mixed reality (MR) is an extension of AR, in which users can interact with digital objects placed in the real world—like shaking hands with a holographic friend placed into a room via an AR headset (Marr 2019). One could argue that these new devices are just a repackaged, modernized form of the same stuff that has been around for decades: movies and telecommunication. Yet the sensorial experience of that which is not, the presence of those who are absent, is becoming ever more 'real.' From visual over audio to tactile and even olfactory, as our senses are addressed in an ever more enthralling manner. While the attraction of our readily available virtual 'reality' increases, the desire to engage with an offline environment that is out of control decreases. Ironically, the fear that many indigenous people have of photography as device to catch their soul (Shanks 2015) is coming to haunt those who consider the Western World as the height of 'civilization.' *Reality* is turning into a fuzzy term where tangible and intangible, material and immaterial features merge. The transition from touchable to un-embodied things is moving ahead at high-speed.[15]

A topic that acutely illustrates the fading transition from technology to terror is gene editing. What seemed once like science fiction is now happening in laboratories around the world.

In 2015, researchers conducted the first experiments using CRISPR (a methodology for gene editing) to edit human embryos. Since then, a handful of teams around the world have begun to explore the process. The promise is to prevent genetic diseases. Yet the approach is controversial because it creates a permanent change to the genome that can be passed down for generations. Furthermore, the line is fine between repairing damage and designing the desirable. In July 2020, experiments that applied the gene-editing tool CRISPR–Cas9 to modify human embryos revealed how the process can make large, unwanted changes to the genome itself, as well as near the target site. This discovery officially stopped further pursuit. "If human embryo editing for reproductive purposes or germline editing were space flight, the new data are the equivalent of having the rocket explode at the launch pad before take-off" (Urnov in Ledford 2020b). Yet it remains to be seen for how long this pause will last, and who will give the greenlight to move ahead.

It seems implausible that the ask for a public moratorium would lead to global consensus on the criteria that would define the line between desirable and despicable. However, the consultation process leading up to such public voting could generate questions and thinking, pulling the topic out of laboratories that operate under the radar. The influence of humans on nature is ever less determined by capability and ever more by ethics. Choosing what *can* be done requires a firm anchor in values and collectively agreed criteria of 'progress.' We are on a road of no return, scratching along the ethical limitations of the existing moral standards (World Economic Forum 2020).

2.4.1.2   Human—Machine Transitions

When we think about the connection, approximation, and potential merger of the human body with technology, a wide range of issues are to be considered. From enhanced human capacities, over the cohabitation of man and machine, to the overriding prioritization of progress in the name of science or business which may lead to the annexation of humanity by robots. A part of technology's promise is that it will enable us to exceed our natural capabilities. One of the areas where that promise is most apparent are brain-machine interfaces (BMIs). Implanted into the brain, they detect and decode neural signals to control computers or machinery by thought. An illustration of BMIs' potential came in October 2019 when a paralyzed patient used one to control an exoskeleton that enabled him to walk. For the time being, the involved costs prevent an application at scale; yet this is the beginning.

The advent of smartphones and wearable electronics has augmented our abilities in ways that would have seemed like science fiction 40 years ago. "We slip on our wristbands and smart-watches and augmented-reality headsets, tuck our increasingly powerful smartphones into our pockets, and off we go—the world's knowledge a voice-command away, our body-metrics and daily activity displayable with a few button-taps" (Kolakowski 2014).[16] So ubiquitous have these types of technology become that we consider them as normal while categorically refuting the idea of being subject to robotic enhancement. Cyborgs are no science-fiction.

Coined in the context of a NASA report, the term 'Cyborg' (Clynes and Kline 1960) has moved beyond the realm of space exploration to encompass an expansive mesh that covers the mythological, metaphorical, and technical sphere (Hess 1995). Commentators have largely used the term 'cyborg' to describe what they see as an unprecedented merger between humans and machines, and to express concerns about the ways in which body and brain are increasingly becoming sites of control and commodification (Rinie van Est et al. 2014). Whereas some define a cyborg as "a person whose physiological functioning is aided by or dependent upon a mechanical or electronic device" (Mann and Niedzviecki 2001), others go beyond this perspective of a hybridization of human/machine, to analyze cyborgs more explicitly in terms of augmentation—as a "person whose physical tolerances or capabilities are extended beyond normal human limitations by a machine or other external agency that modifies the body's functioning; an integrated man–machine system" (Oxford English Dictionary 2014). Under either definition, different people fall in different places on the spectrum of pure human (no interference nor use of technology) to partial cyborgization (external use of tech tools such as cellphones or wearables), to integrated technology (pacemaker, robotic prothesis) to consummate cyborgism (inbuilt technology leading to augmented human capabilities—such as the kind of implanted mesh developed by Neuralink, a company founded by Elon Musk in 2017 (see below), that does not merely support natural bodily functioning (i.e., heartbeat/pacemaker), but expands upon these to equip the human with super-human abilities[17]).

Periodically repackaged as a radical idea, the claim that we are, and always have been, cyborgs has been around since the 1990s. In 1998, it was proposed that wherever the human organism is linked with an external entity in a two-way interaction the result "can be seen as a cognitive system in its own right" (Clark and Chalmers 1998). The idea that humans are already cyborgs is met with resistance from those who note that "pointing to something like cell-phone use and saying, 'we're all cyborgs' is not substantially different from pointing to cooking or writing and saying 'we're all cyborgs'" (Sterling 2008). Which is exactly the argument of those whom they oppose. The latter consider the human "tendency toward cognitive hybridization" not as a modern phenomenon. They perceive the evolution of humanity as marked by a series of "mindware upgrades," from the development of speech and counting (even though they do not involve any artificial enhancement), to the production of moveable typefaces and digital encodings (Clark 2003). The futuristic conception of human-machine hybrids amounts then to nothing more than "a disguised vision of (oddly) our own biological nature" (Deventer 2009). The essential takeaway from this debate is that technology can (under certain circumstances) render human beings more capable than they would have been without it. Seeing this potential, and the risk of abuse that is inherent to any asset, humans are well advised to be aware of the opportunities and caveats that are part of the 'black box' that they take so willingly in their lives.

Awareness and acceptance of our changing relationship to technology including the accelerating cyborgization trend may facilitate the development of policies to accommodate that change adequately. Shying away from acknowledging and addressing the transition is not stopping the latter from happening. We can influence technology only if we recognize its implications with an open mind. Before the transition from using technology to being addicted to it has become irreversible. Whatever the outcomes will be, we are accountable.

2.4.2   Mind—Thinking Re-Thought

Computer science defines the study of artificial intelligence (AI) as the study of 'intelligent agents,' including any device that perceives its environment and takes actions to maximize the chances of successfully achieving a certain goal (Poole et al. 1998). This can be further elaborated to characterize AI as "a system's ability to correctly interpret external data, to learn from such data, and to use those learnings to achieve specific goals and tasks through flexible adaptation" (Kaplan and Haenlein 2019).

As seen earlier (Sect. 2.3), Artificial Intelligence increasingly equals Augmented Influence. It allows those who shape it to analyze and systematically affect those who use it. The better we understand what underpins the attitudes and action of an individual or a group, the more effective an (inter)action geared to persuade them is likely to be. Because we can tailor the interaction based on known triggers. We will look at the emotional impact in the next section. However, already on the intellectual side of things, an overview of available tools illustrates why AI represents an ally for institutions and individuals who seek to expand the scope, reach, and intensity of their ability to not only process information about their audiences but influence the latter. Unfortunately, thus far the range of applications is nearly exclusively exploited for commercial purposes and only sporadically to nurture inclusive social change. Within that field, their usage expands quickly as the sophistication of AI increases the quantity of data that is collected, analyzed, and generated; and vice versa.

2.4.2.1 Quantity Enhances Quality

The speed of connectivity, the rapidly growing number of users, and the data they generate accelerate the development of sophisticated AI. The more data algorithms dispose of, the faster they learn, and the more sophisticated they become. Neural networks relate to AI techniques that are modeled after connections in the human brain, which enables them to learn and improve over time. This leads to Deep learning or 'unsupervised 'machine learning.' This next generation of AI equips computers to teach themselves. Deep learning allows machines to perform high-level thought and abstractions, such as image recognition. This leads to ever more advanced marketing due to customer personalization, audience clustering, predictive marketing, and sophisticated brand sentiment analysis (Kaplan and Haenlein 2019). So far, no AI can successfully replicate the full depth and breadth of human skills and cognition, but already machines are superior to humans at select tasks. Think finding directions in an unknown city (GoogleMaps). A well-known commercial use of their current ability is 'Conversational AI,' commonly seen in online chatbots, which use AI to mimic human conversation via online chats. The range of their abilities expands quickly toward Artificial General Intelligence (AGI) which refers to AI with advanced to human-like intelligence levels that entail self-cognition. The promise (and risk) is that soon it will be impossible to distinguish whether your online counterpart is human or not.

2.4.2.2 Quality Increases Quantity

COVID-19 accelerated the shift to contactless interactions, which leads to additional streams of data. As ever more institutions are using hand-eye coordination, finger pressure, hand tremors, navigation patterns, etc., to confirm a user's identity at log-in, their virtual profile becomes ever more refined (Baron 2018). Big Data refers to extremely large data sets that may be analyzed computationally to reveal patterns, trends, and associations, especially relating to human behavior and interactions.[18] The 4Vs commonly attributed to it—volume, variety, velocity, veracity, determine its value—and its value has been equated to gold. Defined as the practice of searching through large amounts of computerized data to find useful patterns or trends 'data mining' was already used in the 1960s (Merriam Webster Dictionary 2021). Today it is a common term in the vocabulary of tech speak. In 2012, every day 2.5 exabytes (2.5×2^{60} bytes) of data were generated (IBM 2013). Between 2013 and 2020, the global data volume increased to circa 4 zettabytes (4×2^{70} bytes). By 2025, there will be an estimated 163 zettabytes of data (Reinsel et al. 2017). Despite the hype, "Big data is like sex among teens. They all talk about it, but no one really knows what it's like" (Herencia in BBVA 2019).

2.4.2.3   Gather and Use

The size of data sets expands rapidly because there are more and more devices to collect it, including wearables, smartphones, aerial (remote sensing), software logs, cameras, microphones, radio-frequency identification readers, and wireless sensor networks (Segaran 2009). These gigantic quantities of information can be analyzed to identify patterns at large scale. Based on such cross-sectorial analysis, visualizations can illustrate principles and connections that are omnipresent yet hidden in plain sight. Drawing on micro-, meso-, macro-, and meta-dimensional data sources, an interdisciplinary analysis may show surprising links between apparently disconnected realities. For example, a simulated comparison between two of the most challenging and complex systems in nature: the cosmic network of galaxies and the network of neuronal cells in the human brain, brought to light tight similarities (Vazza and Feletti 2020). Despite the substantial difference in scale between the two networks (more than 27 orders of magnitude), their quantitative analysis, which sits at the crossroads of cosmology and neurosurgery, suggests that diverse physical processes can build structures characterized by similar levels of complexity and self-organization. [...] "Structural parameters have identified unexpected agreement levels. These two complex networks show more similarities than those shared between the cosmic web and a galaxy or a neuronal network and the inside of a neuronal body" (Feletti).

We can address only what we are aware of, and seeing is better than blind believing. Visualizing interplays within the *mmmm-matrix* may facilitate intra-dimensional OPTIMIZATION at large scale. Due to the immense information load Big data processes are important in this endeavor. Maximizing the benefit of this latent potential for the common good requires a decoupling from commercial interest. At present, the majority of data processing power is in the hands of private sector players (i.e., FAAMG—Facebook, Apple, Amazon, Microsoft, and Google [now called Alphabet]). In the best-case scenario, they will change track and expand their scope from increasing their power and money to sharing both. Alternatively, a new stream of online power players must rise to shine; these may be forward thinking governments that invest beyond borders—philanthropic individuals or nerds and coders who feel the need for a new WHY. Ideally those three groups unite behind a shared goal.

The separation of data collection and data usage; of information, influence, and manipulation, is shifting subtly. On the one hand, the gigantic quantity of anonymous user data collected by companies on a constant basis provides scientists with tools for in-depth studies that would have been unthinkable in the past: from the psychology of morality, to how misinformation spreads, to the factors that make some artists more successful than others (Ledford 2020a). On the other hand, that abundance of information about people offers insights which can be used to not only understand but influence them. In many instances, users pay for supposedly free services—think Google, Facebook, Twitter, etc.—with information that enables these companies to sell ever more in an ever less resistible manner. We accept and actively support this de facto exploitation of ourselves. We do it because we *want* to. No one forces us to spend hours on Social Media, sprinkling our data all over the web. Or, we go along because it is the way of least resistance; inertia is at play. Awareness of the constant virtual influence on our minds is the first step to protect us from it.

Our brains evolved for specific environments. Artificial intelligence has opened unimagined possibilities to speed up and refine millennia of biological evolution. Among those who want AI to not just support humans but upgrade them is Elon Musk who suggested that humans might not only benefit from "some sort of merger of biological intelligence and machine intelligence", but that they will need it to survive. (Science Focus 2020). He thus founded Neuralink which seeks among other purposes to develop an injectable mesh that connects the brain directly to computers. Such neural lace and other AI-based enhancements are supposed to allow data from the brain to travel wirelessly to one's digital devices or to the cloud, where massive computing power is available. Technological singularity is a hypothetical point in time at which technological growth becomes uncontrollable and irreversible, resulting in unforeseeable changes to human civilization (Eden and Moore 2012). Are we putting the moral foundations of humanity at risk to overcome the limitations of our physiological human framework?

Beyond protecting and harnessing individual willpower to resist exploitation, can technology serve to nurture expressions of generosity and compassion? Can it serve to induce the same level of desire that we experience for technological gadgets, for scrolling, clicking, and liking, regarding the common good?

2.4.3   Heart—The Oxymoron of Artificial Emotions

Artificial Emotional Intelligence (AEI) or Affective Computing is a subset of broader AI which measures human emotion, understands stimuli, and offers feedback. Computer vision identifies facial expressions, and machine learning predicts the underlying emotions based on micro-movements that are imperceptible to the human eye. i.e. They measure stress or anger with the help of sensors to understand the increased blood pressure of the person. Their abilities derive from and support the type of data gathering and machine learning seen above. Passive sensors harvest information about the physical state of the person. This data is then mapped to the cues that may help to interpret emotions in others. Yet it is tricky to decide upon a person's mood, thought, and character based on such artificial interpretations. Because not everyone expresses emotions the same way. Consequently, the use of this type of AI to determine the culpability of accused individuals, or their mental sanity, has repeatedly led to dramatic misjudgments (Barrett 2009).

Despite its limitations, the use of facial recognition as a surveillance tool is rising fast. Regardless of protests by researchers and civil-liberty advocates who have since 2019 raised major legal and ethical concerns (Roussi 2020). These voices of concern are supported by computer scientists who ever so often flag examples of inaccuracies and racial biases in facial recognition AI. Legal challenges have emerged in Europe and parts of the United States, where critics of this type of technology have filed lawsuits to prevent its use in policing. Many opponents point to the surveillance in China's Xinjiang province as an example of how it can be used to limit freedoms. In spite of these worries, usage increases every month amid private and public entities.

On the positive side, AIE could serve to induce and/or nurture pro-social, compassionate emotions, as a conductor to support a behavior that reflects these emotions. In that context, three arenas of AIE are particularly interesting (emotion-recognition, emotion-generation, and augmentation) and needed to reach a new emotionally intelligent epoch of AI (Schuller and Schuller 2018). A word on each:

Recognition—Although recognizing emotions might seem like a uniquely human strength, emotions can be distilled to sets of signals that are measurable like any other phenomenon. The power of a multi-modal approach for emotion-recognition suggests that computers even have an edge over humans. "Eventually computers may surpass humans in emotion detection because of their bandwidth advantage" (Kraus in Krakovsky 2020). For humans, less is more due to our limited ability to take in a large array of dispersed sensorial inputs. Computational advantage also comes to play in recognizing micro-expressions, involuntary fleeting facial expressions that are nearly impossible to spot for the human eye (which renders gut instincts precious). Micro-expressions flash for less than a fifth of a second, requiring the use of high-speed video recordings to capture and new techniques to model them (Tzirakis 2020).

Generation—Responding appropriately to an emotion that has been recognized is more complex. Dialogue systems still follow mostly hand-crafted rules (Schuller and Schuller 2018).[19] However, a new generation of robots is in the making that can respond intelligently enough to influence human behavior in positive ways. Herein one of the biggest obstacles is the need for context: emotions cannot be understood in isolation. Like speech which is hard to make sense of just based on sounds, emotions are too ambiguous to decipher as isolated physiological expressions.

Given people's tendency to mask their emotions, information from one single channel, such as the face, can be misleading. Thus, a more accurate picture emerges by piecing together multiple modalities; lie detectors were an early version of translating invisible physiological signals such as the rate of the heartbeat to make an interpretation of the mind (Jaques et al. 2016). The latest algorithms facilitate multimodal processing, or the integration of signals from multiple channels such as facial expressions, body language, tone of voice, and physiological signals like heart rate and galvanic skin response. The congruence of those channels is telling. Getting the necessary context requires massive amounts of data, which computers have only recently been able to process, thanks to ever more powerful deep-learning algorithms. End-to-end learning means a neural network can use just the raw material (such as audio or a social media feed) and label the results. Involving minimal labeling by humans, this enables machines to learn all by themselves to recognize emotion (Schuller and Schuller 2018).

Enhancement—Sensitivity (theoretically) enables humans to treat others with the appropriate level of deference. Technology can create computer systems that pick up on signals of bravado (i.e., body posture, speech volume) and other emotional signals and then pull them together to generate a reaction that is not just like an average human—who may be moody, angry, unpleasant, selfish, greedy, etc.; but to generate actions that manifest the best that human beings could come up with if they were a full representation of the values that they believe in and aspire to. AI/AIE may serve to identify what drives the individual, bringing to the surface the connection between a person's aspirations and their action, and to systematically trigger and nurture that connection.

The line is fine between betterment and exploitation. Applying AEI to offer everyone the level of care and compassion which they require is a noble objective. But most applications of AIE technologies are not. Systemic influence on emotions is (still) less advanced than detecting them, but the point at the end of the day is manipulation. Human customers must be aware of the ever more pervasive use of AEI, including the (commercial) intentions behind it. This requires acute vigilance because technology slips into our interactions under the radar. Distinguishing real from fake counterparts is becoming hard.[20] To train our awareness abilities we can turn this 'spotting exercise' into a sport to train our own watchfulness.

Despite their enhancement, emotions remain (for now) one feature that distinguishes humans from machines. "Intelligence isn't unique to humans and machines are logically-based systems, but emotion is unique, and it isn't possible to train machines for empathy—at least not yet" (Koenig in Brower 2020). The other feature is our aspiration to purpose.

2.4.4   Soul—Meaning-Making Machines

The purpose of technology is set through coding, from the initial design process that determines the user experience down to the software itself. Top-level coding determines *how* the technology carries out its function. What we sometimes fail to see is the deeper coding. The 'code within the code' sets the goals (the WHY)—the technology's purpose. Understanding this deeper coding begins with a grasp on the mission and vision of the entities that create and commercialize the technologies that we rely on.

In 2018, revelations about the behind-the-scenes of Facebook called the attention of individuals and institutions (Naugthon 2018). As citizens were panicking about data privacy, and researchers were raising alarm bells regarding the manipulative abuse of artificial intelligence, policymakers took a renewed interest in breaking up big tech. This endeavour was short-lived (Pardes 2018), as the subsequent years came to prove. The wave passed, and the tech industry kept on expanding. In 2020, the CEOs of the Big Five—Amazon, Jeff Bezos; Google, Sundar Pichai; Facebook, Mark Zuckerberg, and Apple, Tim Cook, whose combined worth is $265.8 billion—appeared before the US Congress to defend their business practices.[21] Meanwhile, former Uber executive Anthony Levandowski was sentenced to 18 months in jail for stealing trade secrets (Kang et al. 2020). It remains open whether this is just another wave that will fade and, if leave in its wake further expansion of the influence that a few men yield over billions of people around the world.

To be fair, the founders of today's tech giants did not expect their creations to take on a life on their own. The Frankenstein moment must have come as a rude wake-up for all of them. Mark Zuckerberg did not realize, back when he launched Facebook from his Harvard dorm room, that it would grow to become a home for algorithmic propaganda and filter bubbles. YouTube did not expect to become a conspiracy theorists' highlight reel, and Twitter had not anticipated the state-sponsored trolling or hate speech that came to define its platform. Still, their underpinning aspiration was not to serve the common good either. Profit was the name of the game they had come to play.

But technology does not have to be an alien threat to humanity. Whether it is, depends on the human will to tailor and use the available tools with intentions that are informed by the holistic perspective that is presented here, and guided by the values that underpin it.

Humans are the bridge between the natural and the artificial. Through our partnership with machines, we are creating a synthetic intelligence that is part human and part machine. Indeed, our distributed expanding digital networks of communication are rewiring and re-patterning human consciousness through interconnections. Our internal and globalized connectivity transcends cultural and national borders—either forcing us to re-consider our identity and values, or leading to ever more enclosed communities, by reinforcing the us versus them perception. Digging deep into our biological human genome, our natural coding, "technology could be even be perceived as an extension of life itself" (Kingsley 2011).

Building on the previous definition of Technological Singularity, the latter can be perceived at the same time as an existential threat and an existential opportunity for humanity to transcend its limitations. Augmented Humanity (AH) is possible but it requires courageous choices. 'Enhancement,' as outlined above, derives from human intelligence. When it happens, the subsequent transition from artificial intelligence to super-intelligent AI could be very rapid (Shanahan 2015). We must start to contemplate now what this would/will entail in terms of resources, risks, rights, and responsibilities.

Imagine for a moment putting all the above together into one gigantic global matrix that is geared toward inclusive change for the OPTIMIZATION of everyone's inputs and outcomes. Two baskets are to be filled and balanced. One basket to consolidate all available resources—material and immaterial, hard and soft; from natural resources over manpower, from money to knowledge; independently from who owns them, where and why. In the other basket, we lay all the needs, from purely material ones like access to food, water, and internet, to more complex ones like education, health care, and safety, to even more complex ones like avoiding violence and isolation, while nurturing intrapersonal relationships. Already in the 1970s–1980s, economists used Social Accounting Matrixes to represent economic flows to address social needs.[22] They limited their scope to the material side of the global equation. However, an optimization of the *mmmm-matrix* goes beyond economic flows, to embrace the twice 4 dimensions and their ongoing interplays. Big data allows us now to draw information via a vast range of devices that are all part of the technological eco-system. The results may serve to design and constantly refine an organically evolving planning matrix that embraces matter and mind.[23]

2.4.4.1   Riding the Curve of Challenges

The dark side of our infatuation with technology is its all-pervasive influence, and the fact that we are admitting it into our life without much questioning.

The uneasiness that we feel in the awareness of our dwindling control comes from the same place as the unanswered question of meaning. From work to entertainment, we use the internet to participate and to proselytize, to socialize and to separate. It numbs our senses and thoughts, helping us to forget that there is more to life than the shallowness of being perceived. Along the same lines, we tend to forget that technology without humans is not possible, whereas the other way around, life has existed for millennia.

Synthetic biology[24] appears like an oxymoron yet upon closer inspection it illustrates the ongoing merger of flesh and flash. "A fleshy future is one that does recognize all those interconnections and the human realities of technology. But it also recognizes the incredible power of biology, its resilience and sustainability, its ability to heal and grow and adapt. Values that are so necessary for the visions of the futures that we can have today. Technology will shape that future, but humans make technology. How we decide what that future will be is up to all of us" (Agapakis 2020). We must not just look but see and seize the challenges and the opportunities that derive from them.

Investment flows where the intentions are. "We put enormous resources, in the trillions of dollars, into the science and engineering of the cell phone, but not into the science and engineering of how one person can improve the life of another. And I think that's what we're going to need in spades. Because having material abundance is one thing. Having a rich and fulfilling life is another" (Russel 2020).

2.4.4.2   *From Self-Interest to the Common Good*

That material needs and material resources correlate with each other is not new, and as stated already by Mahatma Gandhi "The world has enough for everyone's need, but not enough for everyone's *greed*." What makes the twenty-first century different is that we now have tools to (1) disconnect the pursuit of a better world from the necessarily flawed human desire of outsmarting their own interests and (2) nurture in individuals the understanding that they are part of a global community whose evolution depends on each member. Deployed based on the aspiration to optimize outcomes, algorithms could be designed to come up with practical, unbiased propositions that address the needs based on the available resources; immune to a central character trait of the human species, greed. It may sound like an impossible task but this should not discourage us, because "there's a danger of missing out on something useful just by assuming that it's not possible" (Zlosnik in Wood 2020).

Imagine a software that is systematically attuned to trigger emotions that are conducive to solidarity, such as generosity and compassion. Thanks to machine learning, the underpinning algorithms would get ever better at discerning what influences our feelings, deriving from it our decisions and ultimately our behavior in each situation. Envision a type of technology that is designed with the aspiration to help users better themselves (Fig. 2.1). The line between influence and manipulation is fine. However, Society is marked by centuries of inequity, violence, environmental pollution, and in-group/out-group fighting. This leads to the conclusion that it is challenging for humankind to move alone from an

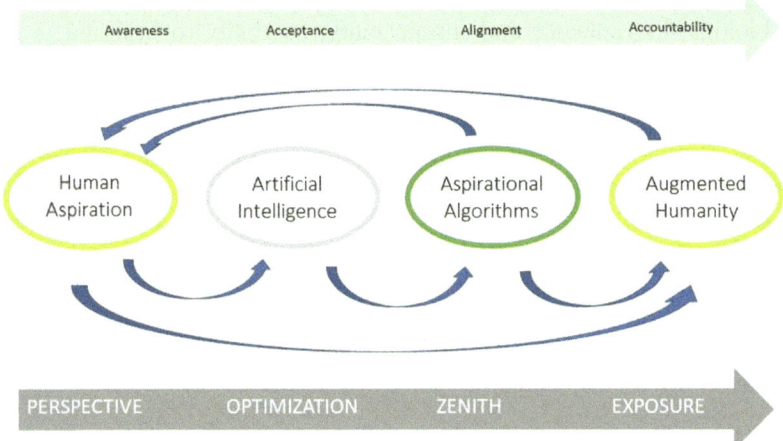

Fig. 2.1 Schema of the possible Transition from human aspiration to augmented humanity. Human Aspirations (HA) are the point of departure for technology that is conducive to an inclusive pro-social society. They underpin Artificial Intelligence (AI), which may become the prototype for Aspirational Algorithms (AA). AA benefit from machine learning as does AI, yet they are intended exclusively to (1) nurture a compassionate generous attitude among users, and hereby (2) increase the quality of life for everyone. Thereby AA accelerate progress toward Augmented Humanity (AH) or a state where a large and ever growing number of people cultivates the common good. The development of AA is conditioned by the evolution of the human mindset, from *Awareness* of interconnected change over *Acceptance* of that all-pervasive continuum; from *Alignment* of the multiple dimensions, to *Accountability* for the efforts that are invested to make the best of them for everyone. Such a mindset benefits from, and is beneficial to, a fresh PERSPECTIVE on offline and online life, the OPTIMIZATION that derives from it, progress toward one's personal ZENITH and as a result EXPOSURE of one's own best self to society (This will be further explored in the concluding chapter)

attitude of 'me' to the aspiration of 'we.' Suppose people have the overarching value orientation of being generous, compassionate, honest and courageous, and genuinely want to reflect this in their behavior. However, the instinctual triggers of everyday life keep on hijacking this aspiration. Could the use of priming technology be justified? Imagine the person decides to put the related app on their phone. Similarly, to millions of users who put screen-locks on their phones to (voluntarily) limit the time they spend on their devices (time of which they objectively know it is de facto wasted and keep on indulging in).

****

"Our version 1.0 biological bodies are likewise frail and subject to a myriad of failure modes. While human intelligence is sometimes capable of soaring in its creativity and expressiveness, much human thought is derivative, petty and circumscribed. Technological Singularity will allow us to transcend these limitations of our biological bodies and brains" (Kurzweil 2005). We will see in the subsequent chapters how hardware (infrastructure) can be shaped in a vision of inclusive, sustainable resource usage; and how software (knowledge) may be instrumentalized to maximize the naturally limited resources of Planet Earth to cover the vital needs of the growing world population.

NOTES

1. Stemming from the Greek 'tekhnologia' it was used in the early seventeenth century to describe 'systematic treatment' (from tekhnē 'art, craft' and -logia).

2. A definition that remains in use since the 1930s refers to technology as a field that "includes all tools, machines, utensils, weapons, instruments, housing, clothing, communicating and transporting devices and the skills by which we produce and use them" (Bain 1937). Whereas scientists and engineers usually prefer to define technology as applied science, rather than as the things that people make and use; this definition is too narrow for those who consider it as any "creative process involving human ingenuity" (Areff 2018).

3. Human rights are rights inherent to all human beings, regardless of race, sex, nationality, ethnicity, language, religion, or any other status. Human rights include the right to life and liberty, freedom from slavery and torture, freedom of opinion and expression, the right to work and education, and many more. Everyone is entitled to these rights, without discrimination. For a detailed overview of rights, see the Human Rights website of the UN.

4. Baby Boomers grew up as television expanded dramatically, changing their lifestyles and connection to the world in fundamental ways. Generation X grew up as the computer revolution was taking hold, and Millennials came of age during the internet explosion. For Generation Z, the above have been part of their lives from the start. Social media, constant connectivity, and on-demand entertainment and communication are innovations

Millennials adapted to as they came of age. For those born after 1996, these are largely assumed (Dimock 2019).

5. While these figures give an indication they must be taken with caution since double counting may be at play, with many users who access the internet via mobile also access via laptops.

6. The sorcerer's apprentice comes to mind: "Herr, die Not ist groß! Die ich rief, die Geister, Werd ich nun nicht los" (Goethe 1797). "The spirits, whom I've careless raised, are spellbound to my power not." We fail to deal with the tools that we created—because we created them with a self-centered mindset, without an all-embracing perspective on the interplaying context.

7. Cyberbullying is a form of harassment in digital communication mediums, such as text messages, internet forums, chat rooms, and social media. As opposed to real-life bullying, online bullying takes advantage of the anonymity of the internet, as well as the possibility to quickly spread rumors, gossip, photos, or (mis)information to large groups of people (Johnson 2021).

8. Deep Blue can identify the pieces on a chess board and know how each moves. It can predict what moves might be next for itself and its opponent. And it can choose the most optimal moves among the possibilities. It has information of past games in its program (i.e., the match it lost to Kasparov) but it doesn't have any concept of the past—or future (Finley 2012) Finley, K. (2012) "Did a Computer Bug Help Deep Blue Beat Kasparov?", *WIRED Magazine*, https://www.wired.com/2012/09/deep-blue-computer-bug/. Accessed November 27, 2020.

9. Representations are internal models of the environment that can provide guidance to a behaving agent, even in the absence of sensory information. It is not clear how representations are developed and whether they are necessary or even essential for intelligent behavior (Marstaller et al. 2013).

10. When people make decisions, they do so against a background consisting of choice architecture (Thaler and Sunstein 2008). Whether it is a cafeteria, a shopping mall or a website, the design affects what people choose. This design can be shaped intentionally to nudge certain decisions rather than others. Nudges are interventions that steer people in particular directions but that also allow them to go their own way. A reminder is a nudge, so is a warning. To qualify as a nudge, an intervention must not impose significant material incentives (Sunstein 2002). For the future, one may imagine new forms of choice architecture that are designed to reduce poverty and environmental pollution; save energy and increase retirement savings; boost the efficiency of social security programs and of healthcare or education settings. Choice architecture may serve to combat discrimination and promote economic growth. AI is a tool that offers the potential of taking such architecture to scale.

11. This kind of complexity is far from science fiction: Uber's experimental platform already runs over 1000 experiments concurrently to adapt their platform experience to the behaviour of small consumer segments. The tools of data science allow us all to embrace this complexity and use it to influence.

12. For example, placing the picture of your mother on your desk may prime you to work harder to achieve your goal, without you being actual aware of this connection.

13. In the US, the concept of 'cultural canceling' was popularized in recent years as a way of demanding greater accountability from public figures who have committed or are accused of having committed some disqualifying moral transgression. It is a (tacit) agreement to not amplify, support, fund entertainers who transgress (some) social norms (Nakamura in Bromwich 2018). On the positive side, it enables those with little political power to litigate perceived injustices in the more accessible forum of popular culture. On the other hand, it entails the risk that a predetermined point of view is being imposed by force of public shaming instead of persuasion. Political correctness is a term used to describe language, policies, or measures that are intended to avoid offense or disadvantage to members of particular groups in society (Merriam Webster). The modern pejorative usage of the term emerged from conservative criticism in the late twentieth century with an implication that these policies are excessive or unwarranted (Friedman and Narveson 1995). The use of rhetoric as a tool to name and shame, stifle and shuffle arguments about topics that walk the fine line of politics and morality, culture and faith, has been around in various shapes and sizes since the beginnings of society. The internet extended the influence and impact of such ready-made value judgments. The same third-hand opinion is echoed and multiplied until it becomes too vast to question even mentally.

The danger of our constant implicit and explicit ostracism of certain topics is threefold. Firstly, speaking up about uncomfortable issues becomes ever less attractive. When every word, every statement is politically loaded, people refrain from thinking, leave alone, talking about certain topics. Afraid to get it wrong, they pick up the opinions of others that fit their overall political orientation. Secondly, it is an elite-exchangism. Gathering information, forming an opinion, participating in a public debate require, even in times where the internet has expanded the public participation square immensely, three assets—access, appreciation, and time. If liberalism ambitions to become a universalist project, rather than a class one, "then it must ask what material conditions are necessary for empowering all people to fully and freely participate in the debates that shape their lives" (Levitz 2020). Thirdly, the risk of inertia expands exponentially. When it is easy to get pre-chewed opinions online, why do not take the trouble to make up your own mind.

14. The Internet of Things is a system of interrelated computing devices, mechanical and digital machines, objects, animals or people that are provided with unique identifiers and the ability to transfer data over a network without requiring human-to-human or human-to-computer interaction. "It's about networks, it's about devices, and it's about data" (Gorski 2020).

15. The 5th generation of mobile internet connectivity provides super-speedy download and upload speeds as well as more stable connections; which allows the transfer of data that would have previously involved physical supports. And while most countries are struggling to introduce 4G, 6G is already in the making.

16. For a detailed overview of the conceptualization of the cyborgization of society, see Wittes and Chong (2014).

17. For example, Neil Harbisson, a cyborg activist and artist born with a form of extreme colorblindness that limits him to seeing in only black and white, is equipped with an "eyeborg," a device implanted in his head that allows him to "hear" color. See TED (2019).

18. Our increasing ability to store and analyze information has been a gradual evolution (starting with tally sticks around C 18,000 BCE). However, things certainly sped up at the end of the last century, with the invention of digital storage and the internet (Marr 2015). Accordingly, the idea that humanity is creating an ever-expanding body of knowledge ripe for analysis has long been popular in academia. First attempts to quantify the growth rate in the volume of data came about in the 1940s. The "information explosion" was first used in 1941, according to the Oxford English Dictionary. Current usage of the term big data tends to refer to the use of predictive analytics, user behavior analytics, or certain other advanced data analytics methods that extract value from data and seldom to a particular size of data set (Press 2013).

19. A phone help line might be intelligent enough to transfer you to a human operator if it senses you are angry or desperate, but still lacks the sophistication to calm you down by itself. That is changing with the new generation of chatbots.

20. Commonly referred to as the Turing test, the imitation game invented by mathematician Alan Turing in 1950 tests a machine's ability to exhibit intelligent behaviour equivalent to, or indistinguishable from, that of a human (Turing 1950).

21. Amazon was accused of abusing its role as both a retailer and a platform hosting third-party sellers on its marketplace. Apple has been accused of unfairly using its clout over its App Store to block rivals and to force apps to pay high commissions. Rivals have said Facebook has a monopoly in

social networking. Alphabet, the parent company of Google, is dealing with multiple antitrust allegations because of Google's dominance in online advertising, search, and smartphone software (Kang et al. 2020)

22. A social accounting matrix (SAM) represents flows of all economic transactions that take place within an economy (regional or national). It is at the core, a matrix representation of the national accounts for a given country, but can be extended to include non-national accounting flows and created for whole regions or area. SAMs refer to a single year providing a static picture of the economy (Arndt et al. 1997).

23. One illustration of AI potential in that regard is the interaction between food and human health. More than 99% of phytonutrients—the natural chemicals produced by plants—are unknown to science. Startup Brightseed uses a proprietary AI platform called Forager to predict the likelihood that plants have useful natural compounds and the likelihood that those phytonutrients will have specific health benefits. Trained on a vast library of biomedical and plant research, the platform uses AI to make connections between plant ingredients and health effects far faster than any human scientist could alone. Now envision a giga version of the Forager type AI that draws on mmmm data.

24. Synthetic biology is a field of science that involves redesigning organisms for useful purposes by engineering them to have new abilities. Synthetic biology researchers and companies around the world are harnessing the power of nature to solve problems in medicine, manufacturing, and agriculture (National Human Genome Research Initiative 2020).

References

Agapakis, C. (2020). *What happens when biology becomes technology?* TED. https://rb.gy/cnb8dj.

Arbib, J., Seba, T. (2020). *Rethinking humanity.* RethinX. https://rb.gy/kswquo.

Amarasingam, A., & Argentino M. A. (2020). The qanon theory. A security threat in the making. *CTC Sentinel, 13*(7), 37–44.

Areff, S. (2018). *Why human ingenuity needs to Tango with technology.* Retrieved November 2020. https://rb.gy/wamnui.

Arndt, C., Cruz, A., Jensen, H.T., Robinson, S., & Tarp, F. (1997). Social accounting matrices for Mozambique 1994 and 1995. *TMD Discussion Paper* 28, Washington D.C.: International Food Policy Research Institute.

Arthur, W. Brian (2009) *The nature of technology.* New York: Free Press.

Bain, R. (1937). Technology and state government. *American Sociological Review, 2*(6), 860–874. https://doi.org/10.2307/2084365.jstor2084365.

Ball, T. (2003). *The federalist with letters of "Brutus".* Cambridge: Cambridge University Press. ISBN 978-0-521-00121-2.

Bargh, J. A, & Chartrand, T. L. (2000). Studying the mind in the middle: A practical guide to priming and automaticity research. In H. Reis & C. Judd (Eds.), *Handbook of research methods in social psychology* (pp. 1–39). New York, NY: Cambridge University Press.

Baron, J. (2018). *Tech ethics issues we should all be thinking about in 2019.* Forbes.

Barrett, L. F. (2009). The future of psychology: Connecting mind to brain. *Perspectives on Psychological Science, 4*(4), 326–339.

BBVA. (2019). *The five V's of big data.* Retrieved October 2020. https://www. bbva.com/en/five-vs-big-data/.

Bromwich, J. E. (2018). Everyone is canceled. *New York Times.*

Brower, T. (2020). *Forbes: Cool new tech trends that will change the way you work.* https://www.forbes.com/sites/tracybrower/2020/07/12/cool-new-tech-trends-that-will-change-the-way-you-work/.

Carr, C. (2008). *The glass cage: How our computers are changing us* (1st ed.). New York: W. W. Norton.

Chalmers, R. (1966). *The conscious mind. How consciousness relates to matter.* New York: Oxford University Press (Reprint 1997). ISBN: 9780195117899.

Chen, C. W., Aztiria, A., Ben Allouch, S., & Aghajan H. (2011). Understanding the influence of social interactions on individual's behavior pattern in a Work environment. In A. A.Salah & B. Lepri (Eds.), *Human behavior understanding. HBU 2011. Lecture notes in computer science* (Vol. 7065). Springer, Berlin, and Heidelberg. https://doi.org/10.1007/978-3-642-25446-8_16.

Cho, A. (2020). Beyond quantum supremacy: The hunt for useful quantum computers. *Nature, 574,* 505–510.

Cichoka. A. (2020). To counter conspiracy theories, boost well-being. *Nature,* 587, 177. https://doi.org/10.1038/d41586-020-03130-6.

Clark, A. (2003). *Natural-born cyborgs: Minds, technologies, and the future of human intelligence* 3. New York: Oxford University Press.

Clark, A., & Chalmers, D. J. (1998). *The extended mind, 58 analysis 10.* Retrieved May 2020. consc.net/papers/extended.html.

Clement, J. (2020). *Worldwide digital population as of July 2020.* Statista. Retrieved November 2020. https://rb.gy/zrnjoz.

Clynes, M. E., & Kline, N. S. (1960, Sept). Cyborgs and space, astronautics. Reprinted in the *Cyborg handbook* (C. Hables Gray, Ed., 1995). Available at cyberneticzoo.com/wp-content/uploads/2012/01/cyborgs-Astronautics-sep1960.pdf.

Corry, L. (1997). Hermann Minkowski and the postulate of relativity. *Archive for History of Exact Sciences, 51*(4), 273–314. https://doi.org/10.1007/bf0 0518231.

Deventer, V. D. (2009). Cyborg theory and learning. In S. Wheeler (Ed.), *Connected minds, emerging cultures: Cybercultures* in online learning 173. Charlotte, NC: Information Age Publishing.

Dienlin, T., & Johannes, N. (2020). The impact of digital technology use on adolescent well-being. *Dialogues in Clinical Neuroscience, 22*(2), 135–142. https://doi.org/10.31887/DCNS.2020.22.2/tdienlin.

Dimock, M. (2019). *Defining generations: Where millennials end and generation Z begins.* Pew Research Center. Retrieved November 2020 https://www.pew research.org/fact-tank/2019/01/17/where-millennials-end-and-generation-z-begins/.

Doidge, N. (2015). *The brain's way of healing: Remarkable discoveries and recoveries from the frontiers of neuroplasticity.* New York, NY: Viking Press.

Eden, A. H., & Moor, J. H. (2012). *Singularity hypotheses: A scientific and philosophical assessment.* Dordrecht: Springer.

Finley, K. (2012). Did a computer bug help deep blue beat Kasparov? *WIRED Magazine.* Retrieved November 2020. https://www.wired.com/2012/09/deep-blue-computer-bug/.

Fitzpatrick, M., Gill, I., Libarikian, A., Smaje, K., & Zemmel, R. (2020). The digital-led recovery from COVID-19: Five questions for CEOs. *McKinsey.* https://www.mckinsey.com/business-functions/mckinsey-digital/our-insights/the-digital-led-recovery-from-covid-19-five-questions-for-ceos.

Friedman, M., & Narveson, J. (1995). *Political correctness: For and against.* Rowman & Littlefield.

Gladwell, M. (2007). *Blink: The power of thinking without thinking.* Back Bay Books.

Goethe, J. W. (1797). *Der Zauberlehrling.* Retrieved November 2020. https://commons.wikimedia.org/wiki/Category:Der_Zauberlehrling.

Goleman, D. (1998). *Working with emotional intelligence.* New York: Bantam Books.

Gorski, C. (2020, January). *Interview on the IoT at digital catapult in wired.* http://www.wired.co.uk/article/internet-of-things-what-is-explained-iot.

Griffin, A. (2017). The human race has peaked in many areas and will now decline, scientists suggest. The Independent. https://www.independent.co.uk/news/science/humanity-peak-human-race-species-age-life-expectancy-str ength-height-maximum-a8097191.html.

Haidt, J. (2006). *The moral roots of liberals and conservatives.* TED. https://rb.gy/fep1gt.

Harbisson, N. (2019). *On being a cyborg.* TED. https://youtu.be/d_mmwr bDGac.

Harper's Letter. (2020). *A letter on justice and Open debate.*

Harvey, D. (1989). *The condition of postmodernity.* In K. Woodward & J. P. Jones (Eds.), (2008). https://doi.org/10.4135/9781446213742.n14.

Hess, D. J. (1995). On low-tech cyborgs, In the *CyborG handbook*. supra note 13, at 371.

Hintze, A. (2016). *Understanding the four types of artificial intelligence*. Govtech. Retrieved July 2020. https://rb.gy/lbwy5z.

Hughes. T. (2004). *Human-built world. How to think about technology and culture*. Chicago Distribution Center. ISBN: 9780226359342.

Hutt, J. (2016). Why do civilizations collapse? *World Economic Forum*. https://www.weforum.org/agenda/2016/03/why-do-civilizations-collapse/.

IBM What Is Big Data?—Bringing Big Data to the Enterprise. Retrieved August 2013. www.ibm.com.

Ienca, M., Andorno, R. (2017). Towards new human rights in the age of neuroscience and neurotechnology. *Life Sciences Society and Policy, 13*, 5. https://doi.org/10.1186/s40504-017-0050-1.

Jaques, N., Taylor, S., Nosakhare, E., Sano, A., & Picard, R. (2016). Multi-task learning for predicting health, stress, and happiness. In NIPS workshop on machine learning for healthcare. December 2016, Barcelona, Spain. http://affect.media.mit.edu/pdfs/16.Jaques-Taylor-et-al-PredictingHealthStressHappiness.pdf.

Johnson, J. (2021). *Cyber bullying - statistics & facts*. Retrieved March 2021. https://www.statista.com/topics/1809/cyber-bullying/.

Kahneman, D. (2011). *Thinking, fast and slow*. London: Penguin Books.

Kang, C., Nicas, J., McCabe, D. (2020). Amazon, Apple, Facebook and Google prepare for their 'big tobacco moment'. *NYT*. Retrieved Aug 2020. https://rb.gy/aml2ap.

Kaplan, A., & Haenlein, M. (2019). Siri, Siri, in my hand: Who's the fairest in the land? On the interpretations, illustrations, and implications of artificial intelligence. *Business Horizons, 62*(1), 15–25.

Kingsley, D. (2011). *Is technology rewiring our soul?* Huffington Post. Retrieved November 2020. https://www.huffpost.com/entry/technology-spirituality_b_854757.

Khedo, Kavi., Suntoo, Rajen., Elaheebocus, Sheik Mohammad Roushdat Ally., & Mocktoolah, Asslinah. (2013). Impact of online social networking on youth: Case study of mauritius. *Electronic Journal of Information Systems in Developing Countries, 56*. https://doi.org/10.1002/j.1681-4835.2013.tb00400.x.

Kolakowski, N. (2014, January 14). *We're already cyborgs*. Slashdot. slashdot.org/topic/cloud/were-already-cyborgs.

Krakovsky, M. (2020). What everyone got wrong about 'the Long Tail'. The Marker.

Kurzweil, R. (2005). *Singularity is near: When humans transcend biology*. New York: Penguin Books.

Ledford, H. (2020a). How Facebook, Twitter and other data troves are revolutionizing social science. *Nature, 582*(7812), 328–330.

Ledford, H. (2020b). CRISPR gene editing in human embryos wreaks chromosomal mayhem. *Nature, 583*(7814),17–18.

Levitz, E. (2020) 'Defending a free society' requires radically changing this one. *New York Magazine/Intelligencer.*

Lippmann, W. (1915). *The stakes of diplomacy* (p. 51). New York: Henry Holt and Company.

Lucas, R. E. (2007). Adaptation and the set-point model of subjective well-being: Does happiness change after major life events? *Current Directions in Psychological Science, 16*(2), 75–79. https://doi.org/10.1111/j.1467-8721. 2007.00479.x.

Mann, S., & Niedzviecki, H. (2001). *Cyborg: Digital destiny and human possibility in the age of the wearable computer.* Toronto: Doubleday Canada.

Marr, B. (2015). *A brief history of big data everyone should read.* Weforum. Retrieved November 2020. https://rb.gy/pmzix5.

Marr, B. (2019). *The seven technology trends everyone must be ready for in 2020.* Forbes. Retrieved January 2020. https://rb.gy/eidsps.

Marstaller, L., Hintze, A., Adami, C. (2013). The evolution of representation in simple cognitive networks. *Neural Computation, 25*(8): 2079–2107.

Merriam Webster (2021). *Digital Mining.* Retrieved March 2021. https://www. merriam-webster.com/dictionary/data%20mining#h1.

Michalos, A. C. (2014). *Encyclopedia of quality of life and well-being research: Social sciences* (Wellbeing & Quality-of-Life). Springer Dordrecht Heidelberg New York London.

National Human Genome Research Initiative. (2020). *Synthetic biology.* Online https://rb.gy/ifbsnw.

Naughton, J. (2018). Facebook's burnt-out moderators are proof that it is broken. *The Guardian.* Retrieved March 2021. https://www.theguardian. com/commentisfree/2019/jan/06/proof-that-facebook-broken-obvious-from-modus-operandi.

Nioeber, J., & Welsh, P. (2020). *Behavioural data science: Ushering in a new age.* https://rb.gy/nuuyzy.

Oluwaseun, A. (2020). *How decentralization could alleviate data biases in artificial intelligence.* Forbes. Retrieved October 2020. https://rb.gy/g2gsky.

Oxford English Dictionary. (2014). Cyborg.

Pardes, A. (2018). Silicon valley writes a playbook to help avert ethical disasters. *Wired.* Retrieved October 2020. https://www.wired.com/story/ethical-os/.

Perrott, D. (2020). Is applied behavioural science reaching a local maximum? *Medium.*

Philipps, R. (2020). It's time to defund social media. *Wired.*

Poole, D., Mackworth, A., & Goebel, R. (1998). *Computational intelligence: A logical approach*. New York: Oxford University Press. ISBN 978-0-19-510270-3.

Press, G. (2013). *A very short history of big data*. Forbes. Retrieved November 2020. https://rb.gy/xbjeax.

Rainee, L., Anderson, J. (2017). *Code-dependent: Pros and cons of the algorithm age*. Retrieved November 2020. Pew Research Center. https://rb.gy/lne2hp.

Reinsel, D., Gantz, J., & Rydning, J. (2017). *Data age 2025: The evolution of data to life-critical* (PDF). seagate.com.

Rinie van Est, V., van Rerimassie., & Keulen, G. D. (2014). *Intimate technology—the battle for our body and behavior*. Rathenau instituut. Available at https://www.rathenau.nl/uploads/tx_tferathenau/Intimate_Technology_-_the_battle_for_our_body_and_behaviourpdf_01.pdf.

Risdon, C. (2017). Scaling nudges with machine learning. *Behavioral Scientist.*

Rosenberg, N. (2008). *Inside the black box*. New York: Cambridge University Press.

Roser, M., Ortiz-Ospina, E., & Ritchie, H. (2019). *Life expectancy*. Our World in Data. Retrieved November 2020. https://rb.gy/vntvmz.

Roussi, A. (2020). Resisting the rise of facial recognition. *Nature*. Retrieved November 2020. https://rb.gy/0oe76f.

Russel, S. (2020, January). *How to ensure artificial intelligence benefits society: A conversation with Stuart Russell and James Manyika*. McKinsey.

Schuller, D., & Schuller, B. W. (2018, September). The age of artificial emotional intelligence. Computer, *51*(9), 38–46.

Schwab, K. (2016). *The Fourth Industrial Revolution: What it means, how to respond.*

Schwab, K. (2019). *The 4th Industrial Revolution: What it means, how to respond*. WEF. Retrieved October 2020. https://rb.gy/hxcllt.

Science Focus. (2020). *Everything you need to know about Neuralink*. Retrieved August 2020. https://rb.gy/wejw4g.

Segaran, T. (2009). *Programming the semantic web*. O'Reilly.

Shanahan, M. (2015). *The technological singularity*. Cambridge: MIT Press.

Shanks, L. (2015). *Do photographs steal the soul*. Retrieved October 2020. https://www.bigbanglife.org/?p=404.

Sterling, B. (2008). BLDGBLOG enters 2009. *WIRED Magazine*. (Retrieved June 2011). http://www.wired.com/beyond_the_beyond/2008/12/bldgblog-enters/.

Sunstein, C. R. (2002). Thinking about risks. In *Risk and reason: Safety, law, and the environment* (pp. 28–58). Cambridge, UK: Cambridge University Press.

Taleb, N. N. (2007). *The black swan: The impact of the highly improbable*. Random House. ISBN 978-1400063512.

Talluri, B. C., Urai, A. E., Tsetsos, K., Usher, M., & Donner, T. H. (2018). Confirmation bias through selective overweighting of choice-consistent evidence. DOI: 2020. https://doi.org/10.1016/j.cub.2018.07.052.

Tasnim, S., Hossain, M. M., & Mazumder, H. (2020). Impact of rumors and misinformation on COVID-19 in social media. *Journal of Preventive Medicine and Public Health = Yebang Uihakhoe chi, 53*(3), 171–174. https://doi.org/10.3961/jpmph.20.094.

Thaler, R., & Sunstein, C. (2008). *Nudge: Improving decisions about health, wealth, and happiness.* Yale University Press.

Tonon, G. (2015). Encyclopedia of quality of life and well-being research. In *Qualitative studies in quality of life: Methodology and practice.* Cham: Springer. https://doi.org/10.1007/978-3-319-13779-7.

Turing, A. (1950). Computing machinery and intelligence. *Mind, LIX*(236), 433–460. https://doi.org/10.1093/mind/lix.236.433. Issn 0026-4423.

Tzirakis, T., Trigeorgis, G., Nicolaou, M. A., Schuller, B. W, Zafeiriou, S. (2020). End-to-end multimodal emotion recognition using deep neural networks. *IEEE Journal of Selected Topics in Signal Processing, 11*(8), 1301–1309.

Vazza, F., & Feletti, A. (2020). *The quantitative comparison between the neuronal network and the cosmic web.* Frontiers in Physics.

Walther, C. (2014). *Le Droit au Service de l'Enfant.* France: Universite de droit, Aix-Marseille UIII.

Wattles, A. (2013): The Golden Rule. Don't take it literally. *Psychology Today.* Retrieved September 2020. https://rb.gy/g5ft8f.

Weingarten E., Chen Q., McAdams M., Yi J., Hepler J., & Albarracín D. (May 2016). From primed concepts to action: A meta-analysis of the behavioral effects of incidentally presented words. *Psychological Bulletin, 142*(5): 472–97. https://doi.org/10.1037/bul0000030.

Wilkinson, R. G., & Pickett, K. (2009). *The spirit level: Why more equal societies almost always do better.* Italy: Allen Lane.

Wittes, B., & Chong, J. (2014). *Our cyborg future: Law and policy implications.* Brookings Education. https://www.brookings.edu/research/our-cyborg-future-law-and-policy-implications/.

Wood, C. (2020). An alternative to dark matter passes critical test. *Quanta.*

World Economic Forum. (2020). *From optimism to realism.* Retrieved October 2020. https://rb.gy/pn3vmx.

CHAPTER 3

Zeniths

Abstract Alignment of aspiration and action conditions holistic social change. It is the logical derivate of *Awareness*. Once we know how the multiple dimensions of our being impact what we do, it becomes natural to seek their optimization through intentional influence. Alignment is the third of 4 stages to individual change. The complementarity of these 4 stages, which seeks to bring out the complementarity of the 4 dimensions of our inner being, and of the 4 dimensions of the society that we evolve in, ultimately leads the person to their highest self, a ZENITH. The number of people who reach that stage conditions the progress meso-level institutions make toward reaching their collective Zenith. In turn, as more and more meso-entities move closer to their Zeniths, the draw of collective dynamics pulls additional individuals onto the new path. Whereas the two first dynamics involve individuals (micro) and institutions (meso), the structural changes that must occur to ensure sustainability encompasses the State (macro) and Supra-national levels (meta). The future depends on our determination to manifest Augmented Humanity. The latter ensues gradually when a critical mass of individuals live up to their highest selves.

Keywords Zenith · Society · Micro · Meso · Macro · Meta · Alignment

C. C. Walther, *Technology, Social Change and Human Behavior*, https://doi.org/10.1007/978-3-030-70002-7_3

Scientifically speaking, the 'zenith' is an imaginary point directly above a particular location, on the imaginary celestial sphere (Glickman 2000). 'Above' relates herein to the vertical direction (plumb line) opposite to the gravity direction at that location (nadir).[1] In the present context, ZENITH is used as a destination whose representation evolves organically, which is in synch with its original meaning, 'the line above the head.'[2] The ZENITH is meant as a vision to orient the journey. It is a reminder that we have barely scratched the surface of our potential. No matter how high we have come, that stage represents only another step on an infinite ladder (Bohm 1980).

Our reality is characterized by wholeness and non-locality. Therefore, everything is instantaneously connected to everything and the Universe (Bohm 1980). Since the latter is full of implicit information that is waiting to become explicit, we cannot explain the status quo by merely dissecting it in the involved parts. The sum is more than its fragments because the universe is constantly enfolding and unfolding along the continuum of change, and the internal and external interplays that derive from this.[3] What we perceive as reality results from "surface phenomena, explicate forms that have temporarily unfolded out of an underlying implicate order" (Peat on Bohm 2013).

The spiral dynamic that derives from this logic illustrates that our progression toward a point X, which appears as the ZENITH of our understanding at that moment is a mobile target (Fig. 3.1). As we reach it, we are already at the onset of surpassing it. The course of change may appear as a regression; as each acquired peak is succeeded by an unprecedented low, that is being followed by a new high that follows the previous trajectory yet surpasses it. An expanding cyclical path (spiral) is one of the various possible paths to go from one zenith to the next one. Jumps, exponential and asymptotic curves may be involved, yet the 'line above the head' remains always higher than the line of sight. The rise and fall of civilizations demonstrate that no valley is static, and no peak permanent. We must give our best, based on our best knowledge of the situation at any point in time. It is this intention-anchored effort that we are accountable for, not the outcomes that result from it. Understanding the spiralling continuum of high and low, up and down, ZENITH and nadir is a useful perspective, because it visualizes the connection of both (transitory) states. In doing so, it illustrates on the one hand our personal power to give the impetus for a change of curve when we are moving through the nadir stage. On the other hand it is a reminder that however

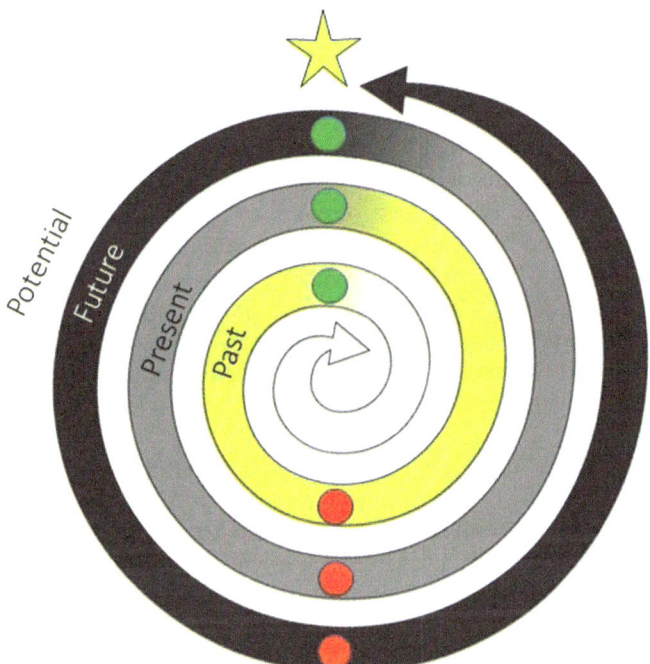

Fig. 3.1 Spiral Zenith versus Nadir. Following a spiral dynamic from the inside towards the periphery we are moving towards a point X, which appears momentously as the ZENITH of our understanding. This apparent culmination is a mobile target. As we reach it, we are already at the onset of surpassing it, regressing seemingly as we move towards an unprecedented low, that is being followed by a new high that follows the previous trajectory yet surpasses it. The 'line above the head' is always higher than the line of sight

pleasant a prevailing ZENITH may be, it will not last—thus attachment to whatever derives from it is futile and prone to pain.

As we progress collectively, we evolve individually. Conversely, through the process of personal betterment, we accelerate the pace of the society that we evolve in. Understood that way, achievement is secondary to pursuit. Similarly, *Alignment* is a process and an outcome. Both are conditioned by the system in which they occur.

In the most general sense, a system is a configuration of connected components, joined within a web of relationships. 'Systems theory' was introduced by Ludwig von Bertalanffy in the middle of the twentieth century.[4] It shifts the emphasis from the distinct parts of an organization to the organization of these parts, while recognizing their interactions not as static and constant, but as dynamic processes. Be it natural like a human being or human-made like a corporation, a system is a cohesive conglomeration of interdependent parts. Each system is delineated by its spatial and temporal boundaries, surrounded and influenced by its environment, described by its structure and purpose or nature, and expressed in its functioning. If it functions in synergy, a system can be more than the accumulation of parts. Changes that affect one component of the system usually affect other components, which in turn affects the entire system. For systems that rely on self-learning and self-adapting, continued performance and growth depend upon how well the system can adjust to its environment.

Systems seek to maintain Homeostasis. The latter is a dynamic state because it derives from constant adjustment to the external changes that the system encounters. Simultaneously to these adjustments the system seeks to preserve an organically evolving stability, by preserving its functionality within a normal range, with some fluctuations required to maintain processes[5] (Martin 2008). Nature, individuals, and institutions are geared toward homeostasis. As mentioned earlier, it takes more energy to put a new dynamic in motion than to pursue a process that is already underway. Establishing new habits by overcoming the beliefs, thoughts, and behavior patterns that we are used to, yet which are outdated, is one illustration of going against a homeostatic situation that is no longer beneficial. That the human being is a 'self-adapting' system can be either helpful or a hindrance when it comes to self-improvement. Pursuing an ideal version of oneself and of society, the ZENITH of life, entails the ongoing OPTIMIZATION of systemic interplays.

Internal synchronization is a decisive factor for the smooth operation of all systems within which humans live, work, and interact. To understand the dynamics of social transformation, we must acknowledge the components of the status quo, including the role of interconnected players and their internal (mis-) alignment. Building on the logic of the *mmmm-matrix*, we look here at aspects that impact the systemic flow that technology has come to (re-) shape. The 4 collective dimensions are at stake—individual (micro), institutional (meso), national (macro), and planetary (meta).

Like *Alignment*, the ZENITH is not a one-off achievement, which can be stashed away once it is reached. Similar to a field of sunflowers in full bloom—even at the peak of its beauty some flowers will be fading already whereas others are just at the cusp, while yet others have perished already. At no point in time will all flowers be at the same point of their respective evolution. Society echoes Nature, and like a sunflower-field, we are a living kaleidoscope. All dimensions matter to reach the state of full blossoming and to extend it in time and space.

For humans and humanity alike, the journey to the ZENITH is work in progress. It is a path that meanders through our existence as long as life goes on. A shift from progress to regress or paralysis can happen anytime. It is a transition.

3.1   MICRO—WHAT DO WE VALUE?

Our self is the transient result of combined elements. Our personality derives from our soul, heart, mind and body and the mutual influence that these dimensions have onto each other and on our surroundings. Every individual is an ever-evolving being. The fluid nature of our Self makes it possible at any time to exit a behavior pattern that is not aligned with our aspirations and values. This section addresses both the 4 individual dimensions (Sect. 1.2) and the role of the individual as part of society (Sect. 1.3). Furthermore, we look at the different layers of attitude toward change that is induced by technology. Building on these pieces we consider the role of individuals in governance, as a way to shape the impact of technology.

Humans evolved to the current state thanks to their intrinsic ability to collaborate (Nowak and Highfield 2011) which is conditioned by the capacity for self-control (Von Hippel 2018).[6] Caring and sharing are part of our DNA (Chapter 1). It is a self-interested instinct that conditions survival. A social set-up in which isolation and bullying are nurtured goes against that inherent set-up. It is thus detrimental—in the short run because it affects individual wellbeing, and in the long term because it jeopardizes collective welfare thus endangering the survival of the species.

At first sight, non-cooperators seem to do better than cooperators and wipe them out. But a set of five complementary mechanisms ensures that natural selection favors cooperation more than selfishness: Direct reciprocity is based on repeated encounters between the same two individuals (I help you and you help me); Indirect reciprocity, based on reputation

(I help you and somebody else helps me); Spatial selection, allowing clusters of cooperators to prevail (neighbors help each other); Group selection occurs if there is competition between groups (members of a group help each other). Finally, Kin selection is based on interactions between close genetic relatives (brothers and sisters help each other).

Society is a composite of individuals. If a critical mass of the latter changes, the overall result changes. The status quo is framed by the values of those who constitute society. Two evolutionary advantages play in our favor when we want to change course-creativity and happiness. The ability to imagine something that does not exist (yet), and the desire of happiness are unique human features (Von Hippel 2018). Using our capacity to think (mind) and feel (emotion), we can envision a different tomorrow and translate this vision into tangible results by choosing our behavior accordingly (body). Happiness is the motivational driver (aspiration) that triggers us into action. "Evolution uses happiness to guide us toward what's in our genes' best interest, giving us the best chance of reproduction" (Von Hippel 2018). It is the pleasant counterpart of purpose orientation that is framed by values.

As seen earlier, values relate to the multi-dimensional interrogation that underpins our individual and collective experiences and expressions. They may be moral, religious, personal, or social—and they do not operate in isolation but exist along a mutually enforcing cultural continuum (Chapter 1). Whereas purpose is the North Star that shows the direction, values are sign-posts along the road.

To create a Society that is hospitable to all requires the repositioning of 'compassionate generosity'—Solidarity, on our respective value scales. Solidarity can be expressed in many ways. A minimum manifestation is the conscious effort to overcome the 'bystander syndrome'.[7] At the extreme end of that scale, such manifestation may resemble 'Eusociality,' a cooperative living arrangement in which members of a species reduce their own chances of reproduction to help raise the offspring of others. It becomes possible if a species follows the above-mentioned steps of 'cooperation' because the members of that species stick together and develop the traits and genes that caused them to cooperate in the first place (Nowak and Highfield 2011). When solidarity is settled high in the individual's perception of value, personal and group-behavior reflect that. Creating a society where people are lifted to reach their potential starts with people themselves.

3.1.1 Under-Utilized Micro-Potential

Every human has a unique set of skills, resources, experiences, person-ality traits, and social connections. This combination enables us to find meaning by fulfilling a purpose that nobody else can accomplish in exactly that manner. Not identifying that 'mission'—the WHY of our being—is the first trap to jeopardize the journey; not acting upon the urge that we feel inside to pursue that Why once we found it is the second. Both barriers can be overcome, once we are aware that they exist. Purpose-oriented Technology (Sect. 4.2.3) can nudge us to take the required steps and remind us to stay on course once we have started, but it cannot walk for us.

Suppose a critical mass of individuals

- agrees that they want to live in a society where generosity, compas-sion, honesty, and courage prevail; and based on this aspiration, they
- re-shape their day-to-day behavior to set an example of what the manifestation of their vision looks like in practice. Now imagine this critical mass of people
- operates within a system that is conducive to their aspiration, with national and supranational mechanisms that seek to elimi-nate inequality, discrimination, and oppression and gets proactively involved if the system does not deliver on these ambitions.

Further and beyond, suppose these people

- design types of technology that reflect their own values, and that are tailored to inspire users to choose the same track.

Furthermore, envision that the ensuing pro-social dynamics

- result in the decentralized development of technological assets that are designed with the stated intention to support, scale, and sustain inclusive social change—and that these assets are fine-tuned with an understanding of the internal individual interplay, and interactions within the *mmmm-matrix*.

In any generation, the aspirations of individuals determined the techno-logical path of the collective. Conversely whether the latent potential not only unfolds but delivers is influenced by the prevailing system, society.

Social systems are human constructs which evolve organically with their human constituents. They may appear immutable. But they are as open and agile as the mindsets that arise and evolve in them. Which conscious-ness prevails depends on all parties and cannot be attributed to abstract and opaque systemic influences. It is a universal Open System and we are all operating with an Open Source technology.[8] Acceptance of this set-up is the condition of genuine progress because it influences the attitude of people to their environment.

Democracies imbue citizens with participation potential, to have a say in who will govern them and thus (in principle) equip them to influ-ence the ideological direction and political roadmap that their country is taking. Like everything else, human-made institutions including democ-racy change. They evolve through time, influenced by the people who create and perpetuate them.[9] Technology can increase citizen-participation and amplify the weight that individual voices have, online and offline; if they are designed, delivered and digested with that inten-tion.

Social media platforms could serve to foster healthy public debate. They affect two fundamental aspects of democracy—deliberation and communication (Sect. 2.3)—which may be positive or detrimental to pro-social participation.

Engineering Facebook & Co to funnel a balanced selection of infor-mation, which is rigorously fact-checked, is possible. The challenge is that we 'like' to hear our own views confirmed, and we enjoy what is colorful or scandalous. Algorithms are tailored to please the user, because user time online equals data which means profitability. "Silicon Valley is an extractive industry. Its resource isn't oil or copper, but data" (Tarnoff 2017).

Companies harvest data by observing our online activity. From activ-ities on social platforms like TikTok, Facebook, Twitter, etc., to Google searches, to more behavioral aspects like how long your mouse hovers in a part of your screen. When these mini actions are paired with those of many others, patterns emerge. These patterns are profitable; for the selling of products, and for advertising—because the more data becomes available the more precise the user profile becomes, including likes and dislikes. This is where intensive influencing starts.

Human brains naturally turn toward more of the same. Spending time with sources of information, online and offline, that confirm our own perspective renders that perspective ever more convincing. People become entrenched in their polarized opinions because based on their own initial choices, algorithms subsequently feed them information that reinforces their existing world view. We are free to go to another source of information, to check another channel or platform, yet the draw of inertia is enticing us to gobble up what is served. Taking in what is put in front of us we feel comforted and confirmed in our existing opinion (which may be wrong) and keep on spreading it within our own social network. This is worsened by the massive spread of rumors and misinformation; whether they are spread intentionally or in good faith. During the COVID-19 pandemic, the word "infodemic" was introduced due to the all-pervasive dissemination of online fake news (Sect. 2.3).

The cycle starts, ends, and is perpetrated by our choices. Twitter started in 2020 to filter what it publishes. It is a small step in the right direction which was helped by the launch of a boycott against Facebook, whereby some major companies cancelled their Facebook ads to protest against the unhindered publication of hate-infused posts.[10] Unexplored micro-potential relates to citizen choices and voices.

We cannot run away from technology. Skipping the grid is an option to step out of technology's influence, but leaving aside practical and logistical caveats, it is self-serving. Escaping lacks the courage to expose our abilities to the challenge of shared growth. A new form of prosperity is at reach if we use technology with the right aspiration. A team of humans and computers, working together, can beat any computer or any human working alone. "Racing with the machine beats racing against the machine. Technology is not destiny" (Brynjolfsson 2013). Beyond the (always available) option to quit the digital space altogether, with all the consequences that derive from this choice in a society where the online world is omnipresent, users can choose what to do with the influence that derives from their presence/absence and behavior, online and offline.

Our attitude toward technology is fundamental to the outcomes of technology on us and others. The 4 stages of our vantage point regarding tools of progress correlate with the 4 parts of the *Scale of Influence* (Chapter 1.4), as each stage corresponds to one dimension of our being.

3.1.2   The 4 Layers of Attitude

Complacency—Do we want technology to disappear from our life? If it was possible (which is unlikely since it does not derive from one central entity but is nurtured and driven by an ever-expanding multi-entity network with countless nodes), would we want it to happen? Suppose humankind stepped away from technology, what would be the consequences in medicine, media, and citizen movement? What we have grown used to as 'normal' would be gone. Few would wish for that. Moreover, and perhaps more importantly, consciously refusing technology's gifts means to renounce on the potential of changing the current trend of inequality towards a dynamic of global goodness. Away from the exploitation and waste, the inequity and abuse derived in the past from a mindset of limitation and egocentrism. Technology can help us to remember our purpose and re-anchor ourselves in values. If it is designed and delivered with that (human) intention. Placed along the *Scale of Influence*, this stage relates to the first step, Inspiration, as it echoes the dormant aspiration for meaning (Fig. 3.2).

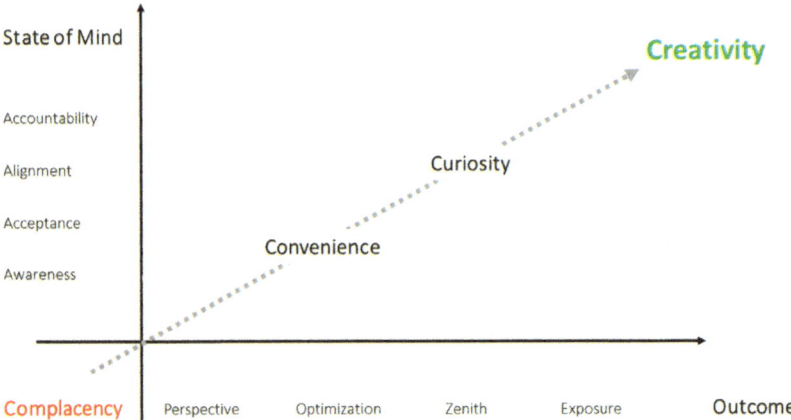

Fig. 3.2   States of mind. From left to right the level of intensity increases, with mindset and outcome connected in a feedback loop of mutual influence. The level of acceptance with which we address technology as part of our present and future is conditioned by the interplay of the 4 stages; whereas the action that derives from each stage influences the others and the overall outcome

Convenience—Do we want technology to expand our comfort-zone, making life ever easier? This was the case of many technological developments in the past—from electricity to telegraphy, from factory automation to cars. The aim that manifested in their design and implementation was pragmatic comfort. Consequences involved less jobs due to more machines, which eventually resulted in new jobs with different profiles. The investments in those technologies paid off as cost-savings in terms of labor and increased production. In the long run, technological progress, political forces, and legal systems led to environmental damage, exploitation, and waste. Everything has a price. On the *Scale of Influence*, this stage relays the second element, Inducement, or the inactivated power to care emotionally about the issues at stake.

Curiosity—Do we want technology to help us become more generous, compassionate, honest, and courageous? Are we ready to shape it as a tool to serve the journey toward a different version of ourselves and the Society that we are part of? Addressed as such it may help us experience reality differently, through improved intellectual abilities, enhanced sensorial intakes, and intensified connections with others. Is our main ambition to know more, do more, have more or to be different, better? The consequences of this choice are irreversible, for better or worse. We are entering a new cycle of modernization, including the elimination of preexisting jobs that will eventually trigger new ones with different skillsets. Gradually, we are moving from the co-habitation of humans and machines to the merger of both, with the integration of machines into humans. This may entail the gradual elimination of social connection and interpersonal warmth or it may bring about a new age of shared space. We can shift from quantity to quality, from being more to being better. On the *Scale of Influence,* this stage conveys the third step, Intrigue, with the awakening interest to not only use but understand and shape the Black Box.

Creativity—Do we want technology to ensure for every human being the highest possible quality of life?[11] Are we ready to use the latent power that derives from knowledge and capacities to translate human rights into humane reality? This choice encompasses a new take on resources, a redistribution of power, and a shift in the roles of players. A radical swing of perspective underpins this 4th choice, because it entails the need to let go some of the positions and privileges that some of us still take for granted. This is the fourth and final stage on the *Scale of Influence* with the Ignition of new ideas, that inspire change, and people to join.

****

Individuals (micro) are humans who are cast into various social roles. Independently of their professional arena, decision-makers remain human beings whose attitudes and actions derive from the same 4 individual dimensions as everyone else. Systematic influence on the 4 collective dimensions is conditioned by the ability to address the individual and professional spheres alike. For the following sections, it is important to remember that those who operate the meso-level (i.e., entrepreneurs), the macro-level (i.e., politicians), and the meta-level (i.e., officials of supranational institutions) have two skins. On the one hand, they are individuals with aspirations, feelings, and beliefs; on the other hand, they are professionals, players, and actors, who pursue their business.

3.2   MESO—WHAT IS THE ROLE OF INSTITUTIONS IN THE PURSUIT OF HUMAN VALUES?

Institutions are strongholds of influence. The latter determines their power and vice versa, their power shapes their influence on the environment they operate in. To optimize society in view of the best possible outcomes for all, individuals and institutions must operate along the same high values.

Institutions are entities established by people, who adhere to "integrated systems of rules that structure social interactions" (Hodgson 2015). The term commonly applies to both informal institutions such as customs, or behavior patterns important to a society, and formal institutions created by entities such as the government and public services. Primary institutions are institutions such as 'the family' that are broad enough to encompass other institutions (Stanford Encyclopedia 2014). Their shape evolves organically in line with the social dynamics of the time, culture, and environment (i.e., the development of the family from multi-generational, to parents with many children, to mono-parenting to same-sex marriages).

Whereas the macro-dimension embraces national governance systems and the overall economy, the meso-level entails the building blocks of the system, including private sector entities. Technological power players have become a primus inter pares in the private sector. Their budget exceeds that of many countries, and their operation outsizes that of any other business sector.[12] Despite that power, which is concentrated in the hands of a few men, they did not endow the world with a solution to face COVID-19. It was not the role of these entities. The question to be pondered is whether it should be.

Should those who hold an influence that derives from gigantic resources be held accountable for the positive or negative impact that derives from the means which they dispose of, on the Society which they operate in—and from which they derive their resources, their influence, and thus their power?

Over the past decade, a form of capitalistic troika emerged to shape the digital economy: Google in search, Facebook in social media, and Amazon in online retail (Solon and Siddiqui 2017). One may add Microsoft and Apple in computer hardware and software. In China, Alibaba took over online retailing, whereas Baidu became the go-to browser. Their power nurtures their power, seeded by consumer data. The more money tech companies make, the more they invest in more proprietary infrastructure such as data centers, and in the refinement of their algorithms. The more customer data they collected thanks to this expanding set-up, the more money they make, which in turn expands their scale and hereby their competitive advantage—to the point where nobody else can keep up.

Even though the size, shape, and scope of institutions varies, they all share one common denominator—they are made up of individuals. And like individuals, institutions are determined by their internal makeup—the why, who, where, and what. Independently of their nature, the answers to these questions are the building blocks of their institutional identity. By consequence, 4 determinants shape their influence, internally and outside—Priorities, People, Positions, Programs.

The *P-Puzzle* is a concept that derives its name from the 4 components (4 Ps) that shape organizations (Walther 2020c). What an organization is and pursues derives from the:

- *Priorities* that justify why it exists,
- *People* who work in it,
- *Positions* that circumscribe how it operates, and
- *Programs* that represent what it does.

The *P-Puzzle* distills the complex nature of institutional influence in components, hereby facilitating the analysis of each component and of the interplay between them, which is necessary for their respective and combined optimization, and hereby the overall best outcome—the institutional ZENITH. The deriving logic is reflected in the following section.

3.2.1   Under-Utilized Meso-Potential

Tech companies reject the idea that they are monopolies, on the basis that customers are free to come and go as they please. And it is true, companies derive their power from customers; consumer choices coronate the kings of the online space. But choice requires (at least) two options. What is the alternative to the internet? During COVID-19, the online space was for many the only lifeline out of isolation; even access to social services and shopping was conditioned in many places by internet access.

In an ideal world, "any institution, whether private or public, exists to serve not only its owners, including those who have contributed even small amounts of capital, but also its workers, customers, suppliers, and the community and environment surrounding it" (Bell 2020). The origins of the social component in for-profit private sector players (businesses) can be traced back to the ancient Roman Laws (Chaffee 2017). By the 1920s, a few CEOs began to assume the responsibility of balancing the maximization of profits with creating and maintaining an equilibrium with the demands of their clients, their labor force, and the community (Carroll 2008). Starting in the 1950s, specific definitions of corporate social responsibilities appeared in the literature (Latapi et al. 2019).

The view emerged that since the businessman's decisions and actions influence their stakeholders, employees, and customers, they have direct impact on the quality of life of society as a whole, and therefore, the responsibility to render that impact positive, or at the very least to compensate for negative impact (Bowen 1953). Corporate Social Responsibility (CSR) began to take hold in the US in the 1970s, when the concept of a "social contract" between business and society was declared by the Committee for Economic Development in 1971 (Chaffee 2017). That social contract was based on the idea that since business functions because of public "consent", it has an obligation to constructively serve the needs of society. Often referred to as "license to operate," such consent (should) compel businesses to contribute more to society than solely the commercial offer of their products and services (ACCP 2017). As it is the case for certain other international instruments, like the International Convention of the Rights of the Child, moral obligations were thus cast in a legal form. This was laudable yet resulted frequently in a de facto downgrade of the underpinning human responsibility that preceded the law (Walther 2014).

Today's understanding of CSR is rooted in the perspective that business entities should go beyond financial goals to consider broader social goals (Korže 2014). It considers that business entities should voluntarily incorporate social, environmental, and ethical standards into their operations to improve the lives of their employees, of the local community, and of society. Unambiguously, the respect for human rights forms thus part of CSR (Čertanec 2019). It should be noted that CSR does not prevent businesses from maximizing profits. Therefore to regulate the space in which corporates operate requires laws and regulations, or, interventions by the State.

Intended to protect individuals, human rights 'secure the moral minimum necessary for us to experience a livable, dignified life as human beings' (Wettstein 2009). When the international human rights regime was set up, states were designated as the sole duty-bearers and the only subjects that could violate international human rights law (UN Economic and Social Council 2006). This has changed over time, and now the subject of international human rights law is deemed 'anyone who is the bearer of rights and duties in international law and is subject to the international legal order' (United Nations Office of the High Commissioner for Human Rights [OHCHR] 2012). No longer only states have obligations in the area of human rights protection, but also non-state actors. Business entities have a responsibility to respect, and proactively protect and promote human rights (UN Global Compact and OHCHR 2017).[13]

The pursuit of CSR is not selfless, since until this day various forms of tax incentives support the social commitment of corporations in many countries. Despite these incentives, and repeated efforts at national and supra-national level to foster and systematize CSR, such as the UN Global Compact—the intentional and disinterested direction of corporate resources to the common good is rare. In many cases, CSR is barely more than a well-intended afterthought, often conceived to embellish the corporation's public image; a decorative layer to cover the ongoing pursuit of profit delivery at all cost. Green-washing and White-washing come in various shapes.

While the simultaneous pursuit of financial and social outcomes may seem counterintuitive, the results are complementary. Businesses with a strong CSR dynamic can boost staff retention and talent recruitment rates (Latter 2017). Staff are happier and generate better operating performance than their peers in less socially responsible companies (Sun and Yu 2015). Their products are in higher demand, especially in times of rising consumer consciousness, and their overall status in the sector's operational environment improves (Latter 2017).

The question of values and socially condoned moral standards is acutely relevant in the conversation of human rights and economic interests, especially when a business's products touch billions of people around the world. The services of the FAAMG influence what people see, think, feel, and do. For social change to be sustainable, their Priorities must Position the wellbeing of People at the center of their Programs. Optimizing the *mmmm-matrix* means to balance the interests of citizens (micro), [tech] corporations (meso), States (macro), and nature (meta). Neither of them operates in a vacuum.

For individuals, it is easy to lose sight of the bigger picture, especially when they are exposed 24 h/7, to the influence of channels that operate with the intention of enticing them to stay and spend online.

What a person wants as a result of commercial titillation is not necessarily aligned with their long-term needs and even their medium-term desires. Facts and features can be filtered, packaged, and polished with the ambition to cover the whole *Scale of influence*—Inspire, Induce, Intrigue, Ignite (Sect. 1.4) Marketing experts are getting ever better at inducing in users the desire for something or another (Emotion). Triggering their curiosity, they intrigue the potential customer to try something new (Mind) and ultimately ignite them to act accordingly (Body). Who is responsible when the deriving choices are harmful to consumers and the society they evolve in? Those who hold that level of influence over individual minds and thus the public space must be held accountable. A checks and balances system that crosses the separation of private and public sector is needed. When public and private power morph into each other, a clear line must be drawn between commercial interest and the common good. A return to the roots of CSR and of human rights is required to instill acute awareness of the universal moral blueprint that preceded them.

Corporations could turn their power to influence the public toward pro-social causes, channeling their all-pervasive presence toward collective benefits without necessarily harming their own profits.

3.3   Macro—What Is the Place of Culture and Legislation in the Upholding of Values?

Human society is a complex interconnected system that must be seen and addressed as such. It encompasses the prevailing models of thought, belief systems, myths, values, abstractions, and conceptual frameworks that help explain how the world works.

Society comprises cultural, political, and economic systems, including the governing structures and institutions that oversee, influence, and manage society and provide the incentive that helps or hinders the decisions, actions, and beliefs of individuals and institutions (Plotkin 2011). In the West, this system entails concepts such as the primacy of empiricism, secular scholarship and scientific method, individual rights, political democracy, nation states, free-market capitalism, and a social contract whereby (most) individuals (are forced to) sell their labor-power to capital. The effectiveness and evolution of one component within this interconnected matrix influence the effectiveness and evolution of the others.

The aforementioned chicken and egg conundrum is at play. Systems can adapt to changing conditions (see discussion about homeostasis at the beginning of this chapter); but their adaptive ability is limited by the speed, scale, and spirit of those who establish and maintain them. When the system is tailored to address not only each component but the interplay between components, the overall outcome improves, as does the outcome of and for each part. Such tailoring requires centralized action and complementarity. Helping individuals capture the possibilities opened by technology is the role of Governments. The State must establish a framework in which every citizen benefits from technological progress. Even though it struggles to adjust to a digitized globalized world, the model of the German Social Market Economy may serve as a positive illustration of tailoring the micro-, macro-, and meso-space to serve the interests of individuals and businesses in a complimentary manner (Abelshauer 2004).[14]

It is a domino-chain that links the micro-, meso-, macro-, and meta-levels in both directions. Micro—The original source of power is are people through their (consumer) choices (votes and voices, clicks and consumption). Meso—Corporate decisions influence what happens at the macro-level which in turn impacts the meta-dimension and the micro-level. Macro—The state has influence on the rules of the market, whereas the market influences which technologies are built and how they are used (Anderson and Rainee 2020). It is arguable whether free markets are essential to innovation, the technological prowess that gave birth to Silicon Valley owes much to government funding and direction (Rotman 2020; Mazzucato 2018).

Over the years, governmental influence has not only declined, a reverse dynamic has begun to operate. Three decades ago, in the early times of the dotcom era, internet companies moved fast, reflecting the understanding that the cyberspace was separate from the physical realm and consequently not subject to the same rules. As encapsulated by the Electronic Frontier Foundation's (EEF) Declaration of the Independence of Cyberspace in 1996: "Governments of the Industrial World, you weary giants of flesh and steel, I come from Cyberspace, the new home of Mind. On behalf of the future, I ask you of the past to leave us alone. You are not welcome among us. You have no sovereignty where we gather" (EFF founding member Barlow).[15]

"We live on a 'platform planet' in which elements of society, like identity, markets and political participation, transcend physical borders" (Saran and Sharma 2019). What is happening online influences what is happening offline. Firms interact with political institutions and citizens through a mix of market and non-market mechanisms. The prevailing political economic system impacts their comparative international advantage (Hall and Soskice 2001).[16] The scale and social consequences of these advantages depend on the State. However, "the public sector's uptake of technology is notoriously sluggish: Government agencies often lag years behind their private-sector counterparts in adopting new digital ways of working" (Gartner 2020). In addition to the difficulty of keeping up with rapidly changing technology, lawmakers are inundated with complaints from their constituents about data use and misuse; their concerns highlight the tradeoffs that from virtually infinite amounts of information at one's fingertips. Not only in the United States "The government enforcement process, when it comes to technology, is inherently behind current technology" (McSweeny in Bloomberg 2020).

Governments are no longer the sole or main gatekeepers of society, but they can and must adjust, to take on the new role for strong centralized action that arises from the present scenario. "No part of this challenge will be ameliorated by American unilateralism, British isolationism or Chinese expansionism – to say nothing of Russian revanchism" (Allen 2019). Increasingly, the influence of States is conditioned by their ability to pursue a fresh vision that is anchored in human values and tailored to fit local needs, national interests, and global circumstances.

3.3.1   Under-Utilized Macro-Potential

While states are polymorphous and poly-contextual, one of their basic functions is to maintain some level of social cohesion (Jessop 2015). The latter is often thought of as the glue that holds society together. While social cohesion is essentially forged by the behaviors and attitudes of individuals, it is an independent quality of social entities, not of individual citizens (Schiefer and van der Noll 2017). Cohesion can be measured at multiple levels (like residential areas, regions or federal states, as well as nation states). Operating along a continuum, social entities can be more or less cohesive (Schiefer and van der Noll 2017). It is an interesting indicator of the interplay of micro-, meso and macro-dimensions, as higher levels of social cohesion were shown to correlate with higher individual levels of agreeableness, conscientiousness, and openness (Larsen et al. 2018). Where a social entity has higher cohesion, this entity will also have inhabitants with a greater prosocial and communal orientation toward others, greater conscientiousness and more openness to experience. Conversely, the prevailing cohesive social environment nurtures certain personality types, including through parental practices and school curricula, which is ultimately conducive to the common good.

Serving the 'common good' (Chapter 1) is one of the State's justifications to be. Since the term 'common good' was first mentioned in ancient Greece, philosophers and economists have differed significantly in their views of what the common good entails and what the state should do to promote it. Rooted in Aristotle's philosophy, a common definition refers to "good proper to, and attainable only by the community, yet individually shared by its members" (Aristotle in Dupré 1993). Despite their disagreements, thinkers through time have stated that the common good is (or at least should be) the end of government; that it relates to the benefit of all citizens, and that no government should become the "perverted servant of special interests" (Diggs 1973), whether these special interests be understood as Aristotle's "interest of the rulers," Locke's "private good," Hume's and Madison's "interested factions," or Rousseau's "particular wills." But even beyond such interests, it may be that well-intentioned politicians do more harm than good in the pursuit of a certain dogma. "Practical men, who believe themselves to be quite exempt from any intellectual influences, are usually the slaves of some defunct economist" (Keynes 1935).

To serve the common good, governments require value orientation on the one hand and neutrality on the other. In reality, public officials tend to listen to those who have money and power. It is a common theme that has meandered through the four industrial revolutions. Whether behind the scenes or in public, the private sector systematically influences national and international appointees. The multi-layered, mutating, and massive manifestations of corporations are driven by the quest for profit. Governments have a choice, when it comes to whom they listen to, and how they deal with those that talk. It is an important choice because "States conceived as organizations claiming control over territories and people may formulate and pursue goals that are not simply reflective of the demands or interests of social groups, classes, or society" (Evans et al. 1985). Governments must step out of the puppet role in which they allowed themselves to be cast by powerful lobbies.

Over the past years, three models emerged to shape the Quadrant of State/Market/TechSector/Citizens—with one power-player in each corner. Though that Quadrant links only selected elements of the *mmmm-matrix*, a fluid intersection of private and public space appears.[17] In China, for instance, a close collaborative relationship between commercial entities (meso) and the political interests of the state (macro) is entertained.[18] In contrast, in Europe, governments (macro) seek to protect individuals (micro) with regulations such as the General Data Protection Regulation (EU GDPR).[19] Despite its weaknesses, the GDPR is a major step toward delimiting the free-for-all approach of tech companies to consumer data. In the United States, the relationship between tech companies (meso) and the government (macro) is marked by minimal regulation, under the guise of "economic liberty" (Marr 2020). This leaves matters in the realm of interaction between citizens/consumers (micro) and entrepreneurs/companies (meso)—who have widely different needs, objectives, and means to defend themselves or pursue their objectives. Within this overall schema, the power of firms themselves varies significantly, distinguishing small and medium and small size enterprises from those that are sufficiently large to operate at the meta-level—i.e., the FAAMG entities.

As seen at the meso-level in the context of CSR, a new logic of checks and balances is needed that does not only limit the power repartition within one institution or even within one *m*-dimension, but across all

4 *m*'s. The shift from business interests to citizen wellbeing and social interests will not happen by itself. Free market forces are shaped for the survival of the fittest. Governments must defend those who are not (yet) fit to fend for their own needs. The principal duty bearer of human rights remains the State. It must defend the fundamental entitlements of its constituents and refrain from infringing on their rights itself.

3.4 META—WHAT POSITION HOLD SUPRANATIONAL INSTITUTIONS, GIVEN THE GLOBAL REACH OF TECHNOLOGY?

The sum is greater than its parts. However, in some circumstances competing agendas lead to the mutual impediment of parts within a system. This has repercussions on the environment in which it operates, on other systems and ultimately the (dysfunctional) system itself. The result is a dynamic which generates ever more unhappy people, dysfunctional organizations, and imbalanced systems.

Society is marked by opposites that are complementary (Chapter 1). The existing interplays can be optimized to improve the outcome for all parties. Nature and technology, old and new power, hardware and software, hard and soft power all drawing on different forms of influence play a role within the kaleidoscopically evolving puzzle of Society. Due to the limitation of resources, redistribution may be required, entailing the rerouting of global resources to address local needs in an equitable way. An attitude of borderless compassion might serve to shape a 'glocalized' logic (global perspective + local application), catalyzed by glocal mindsets (global perspective + local commitment).

The future is likely to be shaped by shifts that are catalyzed by citizen behavior. Maybe change will be like the Renaissance that sprung up in merchant-sponsored prosperous cities like Florence, Bruges, and others that represented the center of wealth at the dawn of the Middle Ages. But more likely, it will be triggered from the bottom-up, via an organically growing distributed network, which is nurtured by creative individuals and groupings around the world (Follows 2017). "Like ink dots on blotting paper, these conscious and creative centers will spread their influence through decentralized channels and processes until a time will come when the ink dots begin to fill the blotting paper" (Kingsley 2011).

The future of power is less about having but sharing. Control, authority, and sovereignty are being reconfigured in a digitally disrupted, disintermediated, and decentralized world. Localization and globalization, connection and isolation are opposing streams that happen simultaneously, at an accelerating pace. Either the old adjusts to the new, or it is replaced by new, globalized authorities and power holders (possibly yet not necessarily for-profit corporations). Those entities would derive their 'mandate' from their de facto influence. The challenge would then become—as it is now the case for State actors—to analyze and address in real-time the voices and needs of their supporters, which determine their power to establish guidelines, regulations, and limits beyond the control of one nation (Kuhel 2017). Shifting the balance between citizens, for-profit corporations, and State players within a holistic perspective of the *mmmm-matrix* has benefits at scale. But where to begin?

Individuals are neglectable quantities if they are isolated. Unless they unite they cannot tip the scale. Businesses seek profit. That is their traditional DNA and the 4th Industrial Revolution will not automatically shift the balance from current power holders to different ones, from firms to citizens. Governments are inclined to concentrate on national agendas, which may be shortsighted and/or influenced by the personal interests of certain individuals, or special interest groups (i.e., tech lobby, oil lobby). The bunkering of protective equipment and vaccines during COVID-19 is the latest illustration of this tendency.

3.4.1   Under-Utilized Meta-Potential

However, there is a large supranational player, dedicated to making the world more just and peaceful; whose vocation it is to help those who cannot help themselves. The non-profit sector, including non-governmental organizations (NGO) and the United Nations (UN) System, is a fertile ground to initiate a fresh global dynamic. The latter may gradually draw in ever more individuals and institutions from other sectors. Primus inter pares among these institutions are the United Nations (UN). A supra-national institution created in 1945 to promote peace and prosperity for the citizens of all nations.

"We the peoples of the United Nations, determined, to save succeeding generations from the scourge of war, which twice in our lifetime has brought untold sorrow to humankind, and to reaffirm faith in fundamental human rights, in the dignity and worth of the human person, in the equal rights of men and women and of nations large and small, and to establish conditions under which justice and respect for the obligations arising from treaties and other sources of international law can be maintained, and to promote social progress and better standards of life in larger freedom, […]" (UN Charter 1945) Dedicated to the pursuit of these goals and the promotion of human rights, the UN appears as an ideal place to start a 5th Revolution. One that is not merely industrial but social, that does not disrupt and destroy, but nudge and nurture, connecting beyond boundaries to optimize the interplay between and within people and peoples.

One example to get started is data management.

Corporations are directly in touch with the individual—exerting the same type of direct influence as do State actors such as schools, hospitals, and law enforcement entities. Increasingly, the hardware and software designed by private sector entities are present at every stage of an individual's life. Paradoxically, as discussed in Sect. 2.1, we are reluctant to share our health data, personal history, banking details, relationships, salary with the Government, with neighbors, and even close friends or relatives. And yet we willingly entrust this information to entities like the FAAMG for the privilege to spend (our own) money and time on their goods. Interfering with traditional strongholds of governments, technology firms are involved in education and employment, cultural offers, and healthcare provision.

Whether used by private or public sector players, "Data collection isn't apolitical" (Achiume in Fallon 2020). Robotic lie detector tests at European airports, eye scans for refugees, and voice-imprinting software for use in asylum applications are among new technologies flagged in a 2020 report of the Special Rapporteur on racism (OHCHR 2020 - A/HRC/44/57). Another one is "bio-surveillance"—focused on tracking people's movements and health metrics, a topic that has become acutely relevant during COVID-19. The implications of these mechanisms on human rights are opaque; and technology increasingly outpaces legal frameworks. This expanding grey zone is either condoned by Governments, or actively supported. "Apart from the insufficiency of legal safeguards, we also lack transparency; and this is not only remarkable, but highly problematic" (Kakavoulis in Fallon 2020). Looking beyond short-term wins for surveillance, there are major ethical concerns regarding the regulation of large-scale data harvesting from populations.

In particular, if such harvesting happens without the consent, or even without the knowledge of the concerned individuals. Since late 2019 a growing number of scientists have been urging researchers to avoid working with governments, firms, and universities linked to unethical projects, to re-evaluate how they collect and distribute facial recognition data sets and to rethink the ethics of their own studies. But their voices are isolated, and not always impartial. Scientists must acknowledge the often morally dubious foundations of science itself—including studies that have collected enormous data sets of images of people's faces without consent, many of which helped to hone commercial or military surveillance algorithms (Noorden 2020).

Who 'owns' all the data that individuals are entrusting to corporations every day? Does it belong to the one who hands it over (customer) or the one who receives it as part of a transaction (corporation); is it a public good that can be accessed and analyzed for collective research interests (academia) or to gather relevant information such as the control of a pandemic by a central institution that represents all of us (government)? What about Big Data initiatives that involve individuals who are not even aware that they are involved, such as the analysis of sewage water in certain neighborhoods to identify the presence of pathogens or drugs (Mallapaty 2020)? There is not one answer to this interrogation—but no country can stay silent and leave the space to the private sector to develop default options (as it is currently the case). Macro- and meta-players must defend the micro-constituents who put them into power and pay their presence and (in)action with their taxes.

A supranational solution to the data conundrum is needed. Because answers involve conscious consumer choices, corporate responsibility, and proactive governments—under the umbrella of a global vision that is anchored in the aspiration of the common good and enforced with cross-border rules and regulations.

Beyond the protective angle of the data dimension, an immense opportunity derives from tackling it on a supranational level.

Artificial intelligence and the ever more sophisticated set of technological tools to gather data grant humanity an analytical power that is far beyond anything previously imaginable. Globalized Big data streams can serve to visualize the mutual streams of influence that shape our multidimensional existence. Taking the pulse of dynamics at macro-, meso-, and micro-level, compounded with the holistic vision of interactions within the *mmmm-matrix* increases the likelihood that policy choices are appropriate. Systemic Big data management can increase the accountability of those who make those choices and improve service delivery by public and private sector entities alike. Situated within a fractal logic, it may thus expand the efficiency and effectiveness of stakeholder interactions. Supranational data flow control can contribute to a global dynamic in which every component is seen and valued. The UN might serve as a guardian to establish and manage such an impartial framework. The impact for society could be hugely beneficial—if the human aspirations that underpin the system and the technology that emerges from it are purpose-oriented (goal) and value-anchored (foundation); if it is free from personal and national interests.

* * * *

There is no miracle cure to the issues that mar society. Humans must shift their own mindset and deriving behavior, *before* a type of technology can be built that makes the pursuit and promotion of such a mindset easier. It sounds tautological but it is simple—what goes in comes out. The design of technology is like planting trees. When we put a potato in the ground, we cannot expect a mango tree to grow.

The Journey to the ZENITH is work in progress, and each step must be walked by individuals, offline in a very traditional manner—by aligning their aspiration for good, with good action. It is worth the effort, not only because a pro-social orientation is conducive to personal wellbeing (Chapter 1), but because a technology that is geared toward inclusive social progress can actually change the World. Out of the 169 targets underpinning the Sustainable Development Goals (SDG), 70% could be enabled by technology applications that are already deployed today. The 4th Industrial Revolution brings a drastic jump of potential (Schwab 2019)[20] and for now, only a fraction of this potential is actively utilized at scale (WEF 2020a). Unlocking latent opportunities requires awareness of the possibility, creative minds to seize it, and courage to follow through on ideas with a holistic PERSPECTIVE.

The year 2020 illustrated once more that systemic problems—a pandemic, a global economic crisis, nationalism, climate change, inequity, racial injustice—require systemic solutions, because they are connected through causes and consequences. *Awareness* of interplays and *Acceptance* of the responsibilities that ensue from them endow us to work on the gradual *Alignment* of the resources at our disposal. We are *accountable* not for the results of these efforts, but for EXPOSING ourselves to the test of our ability in their pursuit.

NOTES

1. At a given location during the course of a day, the Sun reaches not only its zenith but also its nadir, at the antipode of that location 12 hours from solar noon.

2. The word "zenith" derives from an inaccurate reading of the Arabic expression سمت الرأس (samt al-ras), meaning "direction of the head" or "path above the head," by Medieval Latin scribes in the Middle Ages (during the fourteenth century), possibly through Old Spanish (Corominas 1987).

3. "In the enfolded order, space and time are no longer the dominant factors determining the relationships of dependence or independence of different elements. Rather, an entirely different sort of basic connection of elements is possible, from which our ordinary notions of space and time, along with those of separately existent material particles, are abstracted as forms derived from the deeper order. These ordinary notions in fact appear in what is called the 'explicate' or 'unfolded' order, which is a special and distinguished form contained within the general totality of all the implicate orders" (Bohm 1980). On the perspective of 'transcendental realism' which derives from a critique of the dominant positivist and neo-Kantian traditions in the philosophy of science see (Bhaskar 1978).

4. Von Bertalanffy (1968, 1972). The notion of general systems theory first stemmed from the pre-Socratic philosophers and evolved throughout the ages until its formal structuration in the early 1900s. The theory has three main aspects: systems science, or scientific exploration and the theory of systems in various sciences; systems technology or the problems arising in modern technology; and 'systems philosophy'.

5. Homeostasis can break down. The influence of external and internal forces may be such that the homeostatic mechanism cannot cope. For instance, the human body (micro) has several mechanisms to maintain body temperature within a narrow range when the external conditions change (i.e., blood circulation, sweating, shivering). However, when the gap to compensate is too extreme, or the compensation mechanism is activated for too long it becomes impossible to maintain. We can thus die from heat exposure or extreme cold. Similarly, while societies (macro) tend to reproduce themselves in a rather stable fashion, often evolving at an imperceptible pace, at times events upend that pace; external events (i.e., pandemic, natural disaster) or revolutions may occur that drastically

change the social structure, resulting in a different society, even if most of the people before and after the event are the same.

6. Self-control has major social advantages. If I can control the way I respond to you, then I can give you the kind of impression that I want to give you. I can refrain from expressing my thoughts/emotions unless the time is right (Von Hippel 2018). It is an evolutionary advantage that humans have. The level of self-control increases with the understanding of and influence on the dimensions that shape the instinctive drivers of our behavior.

7. The Bystander syndrome comes to play when individuals are less likely to help someone in need when other people are present; usually the greater the number of bystanders, the less likely it is that one of them will help (Darley 1970). Several factors contribute to the bystander effect; ambiguity, group cohesiveness, and the diffusion of responsibility reinforce the individual and mutual tendency to deny a situation's severity. Even though the bystander effect has been considered as one of the most solid and replicable psychological concepts since it was established in the 1970s, recent research gives reason to think that an opposite dynamic might apply in different contexts (Philpot et al. 2019). Once we are aware of tendencies that we are prone to, and acknowledge them, we allow ourselves to address, and prevent, them. *Awareness* comes with *Accountability* for (in)action.

8. Open systems are computer systems that provide some combination of interoperability, portability, and open software standards (Dickinson 1991). Open-source software (OSS) is a type of computer software in which source code is released under a license in which the copyright holder grants users the rights to use, study, change, and distribute the software to anyone and for any purpose (St. Laurent 2008). For decades, technology firms resisted both, as a threat to their market power. Those who conceded eventually survived.

9. A good description of this evolution is found in Held (2006). Democracy could be more than just periodic voting as explained by Dahl (2015). According to Held (1995), democracy must adapt to the new conditions brought forth by the rising power of international actors, including private corporations.

10. In July 2020, Facebook CEO Mark Zuckerberg agreed to meet with the civil rights organizers behind the boycott which had led big brands like Hershey's, Pfizer, and Levi Strauss to cancel their advertisements. Despite the relatively small impact on Facebook's finances whose biggest add spenders were not involved in the boycott, the initiative shook up FB investors and caused a reputational nightmare for Facebook. It is an illustration of the potential power of vigilant users—which can be amplified and expanded upon by conscious engineers and software designers.

11. Quality of life commonly entails health, material comforts, personal safety, relationships, learning, creative expression, opportunity to help and encourage others, participation in public affairs, socializing, and leisure (Michalos 2014).

12. The five tech giants (Facebook, Amazon, Apple, Microsoft, Google) generate a combined US$ 900 billion per year, which equals the annual GDP of some countries among the top 20 economies in terms of total income, i.e., Saudi Arabia or The Netherlands Wallach (2020). For details on the sources of their income, which depends on consumer behavior, see https://www.visualcapitalist.com/how-big-tech-makes-their-billions-2020/.

13. The UN Global Compact is a call to action to business leaders, aiming to mobilize a global movement of sustainable companies and stakeholders to create a world that is conducive to achieving the Sustainable Development Goals (UN Global Compact and OHCHR 2017)

14. Introduced by Ludwig Erhard in 1948, the goal of the Social Market economy is the greatest possible prosperity with the best possible social protection (Erhard 1958). It is about benefiting from the advantages of a free market economy, which includes free choice of workplace, pricing freedom, competition, and a wide range of affordable goods, while at the same time protecting citizens from its disadvantages, such as monopolization, price fixing, and existence-threatening unemployment. This is why the State to a certain extent regulates the market and protects its citizens against illness and unemployment through a network of social insurance schemes. The Social Market Economy was a peculiar and successful embodiment of what came to be known as the Keynesian Welfare State. Like every other (temporary) ZENITH, it had limitations (for instance in terms of gender equality and environmental concerns). Eventually it had to be changed to preserve the positive elements which it had brought about, see inter alia Boyer and Mistral (1978), Jessop (2002), Offe (1984), and Therborn (1987).

15. It that context of techno-liberalism, it is interesting to note that large amounts have gone into lobbying by and for technology companies, e.g., in the US, in 2017, FAAMG spent $49 m on Washington lobbying; oiling a revolving door between Silicon Valley executives and senior government officials in the White House. "They're doing this to protect their oligopolies" (Solon and Siddiqui 2017). Main areas of concern included the threat of action over anti-competitive practices, covering anything that might lead to higher taxation, net neutrality, and privacy. Beyond direct lobby spending which is publicly reported, Silicon Valley exerts influence on policymakers and citizens through "soft power" techniques, including the funding of think tanks, research bodies, and trade associations who lobby the government or stimulate civil society and individual consumers.

16. Hall and Soskice (2001, p. 7) note five distinct spheres of interaction: (1) industrial relations, which entails wage bargaining and concrete workplace conditions; (2) vocational training and education, which involves the resources that firms and workers devote to the development of specific skills; (3) corporate governance, which applies to the availability of finance between firms and investors; (4) inter-firm relationships, which consist of the coordination between the firm and its suppliers, clients, and competitors; and finally (5) employee relationships, which encompass the institutional sociology of the workplace.

17. These constellations of political/legal/market forces are reminiscent of the conceptual contrasts pointed out in the 1990s regarding the State's role in modern capitalism/welfare systems. Esping-Andersen outlined a typology of welfare capitalism in an attempt to classify contemporary Western welfare states as belonging to one of three "worlds of welfare capitalism which are characterized by a specific labor market regime and a specific post-industrial employment trajectory": Liberal regimes (characterized by modest, means-tested assistance, and targeted at low-income, usually working-class recipients), Conservative regimes (typically shaped by traditional family values and prone to encourage family-based assistance dynamics), and Social democratic regimes (universalistic systems that promote an equality of high standards, rather than an equality of minimal needs) (Esping-Andersen 1990). He argued that current economic processes, such as those moving toward a postindustrial order, are shaped not by autonomous market forces, but by the nature of states and state differences. A decade later, Hall and Soskice provided basic operational distinctions between two 'ideal-types' of political economies: liberal market economies (LMEs) and coordinated market economies (CMEs) (Hall and Soskice 2001). Both models loop back to the question of the State's (unfulfilled) potential to shape a minimum social standard of quality of life for all.

18. In an interesting turn of events in November 2020 Beijing proposed drastic anti-monopoly rules for its giant internet companies (Liu and Ren 2020). It is to be seen what comes next.

19. The General Data Protection Regulation (EU) 2016/679 (GDPR) is a regulation in EU law on data protection and privacy in the European Union (EU) and the European Economic Area (EEA). It also addresses the transfer of personal data outside the EU and EEA areas. https://gdpr-info.eu/.

20. The Fourth Industrial Revolution is disrupting almost every industry in every country. Situated in a globalized society, the breadth and depth of these changes herald the transformation of entire systems of production, management, and governance (Schwab 2016).

REFERENCES

Abelshauser, W. (2004). *Deutsche Wirtschaftsgeschichte: seit 1945*. München: C. H. Beck. ISBN 978-3-406-51094-6.

ACCP. (2017). *Corporate social responsibility: A brief history*. https://rb.gy/slwz4c.

Allen, J. R. (2019). World Economic Forum *"Competition or cooperation"*, quoting Brookings John R. Allen in "Disrupting the International Order". https://rb.gy/esfnxf.

Anderson, J., & Rainee, L. (2020). *Many tech experts say digital disruption will hurt democracy*. Pew Research Center. file:///C:/Users/cornelia/Downloads/PI_2020.02.21_future-democracy_REPORT.pdf.

Bell, K. (2020). Quartz weekend editorial. *Business Journalism*.

Bhaskar, R. (1978). *A realist theory of science*. Harvester Press.

Bloomberg. (2020). *Regulation and legislation lag behind constantly evolving technology*. Retrieved November 2020. https://pro.bloomberglaw.com/regulation-and-legislation-lag-behind-technology/.

Bohm, D. (1980). *Wholeness and the implicate order*. New York: Routledge. ISBN 0-203-99515-5.

Bowen, H. R. (1953). *Social responsibilities of the businessman*. Iowa City: University of Iowa Press.

Boyer, R., & Mistral, J. (1978). *Accumulation, inflation, crises*. PUF.

Brynjolfsson, E. (2013). *The key to growth? Race with the machines*. TED. https://go.ted.com/6b7a.

Carroll, A. B. (2008). A history of corporate social responsibility: Concepts and practices. In A. M. Andrew Crane, D. Matten, J. Moon, & D. Siegel (Eds.), *The Oxford handbook of corporate social responsibility* (pp. 19–46). New York: Oxford University Press.

Čertanec, A. (2019). The connection between corporate social responsibility and corporate respect for human rights. *DANUBE: Law, Economics and Social Issues Review, 10*(2), 103–127. https://doi.org/10.2478/danb-2019-0006.

Chaffee, E. C. (2017). The origins of corporate social responsibility. *University of Cincinnati Law Review, 85*, 347–373.

Chopra. D. (2015). *The power of group meditation*.

Corominas, J. (1987). *Breve diccionario etimológico de la lengua castellana* (in Spanish) (3rd ed., p. 144). Madrid. ISBN 978-8-42492-364-8.

Dahl, R. (2015). *On democracy* (2nd ed.). London: Yale University Press.

Darley, J. M., & Latane, B. (1970). *The unresponsive bystander: Why doesn't he help?* New York, NY: Appleton Century Crofts.

Dickinson, I. (1991). *Open systems strategy from IBM*. Retrieved November 2020. Newsgroup: comp.unix.misc.

Diggs, B. J. (1973). The common good as reason for political action. *Ethics, 83*(4), 283–284. https://doi.org/10.1086/291887.

Dupré, L. (1993). The common good and the open society. *The Review of Politics, 55*(4), 687–712. https://doi.org/10.1017/S0034670500018052.

EFF. (1996). *A declaration of the independence of cyberspace*. Retrieved October 2020. https://www.eff.org/cyberspace-independence.

Erhard, L. (1958). *Prosperity through competition*. Thames & Hudson.

Esping-Andersen, G. (1990). *The three worlds of welfare capitalism*. Princeton: Princeton University Press.

Evans, P., Rueschemeyer, D., & Skocpol, T. (1985). *Bringing the state back in*. Social Science Research Council (U.S.). Committee on States and Social Structures, Joint Committee on Latin American Studies, Joint Committee on Western Europe. Cambridge: Cambridge University Press.

Fallon, K. (2020) UN warns of impact of smart borders on refugees: 'Data collection isn't apolitical'. *Guardian*. Retrieved November https://rb.gy/1t60ct.

Follows, T. (2017). The future of power. *Campaign Live*. Retrieved October 2020. https://rb.gy/rmfqht.

Gartner. (2020). *Report: Public sector technology*. Retrieved November 2020. https://www.gartner.com/en/information-technology/trends/public-sector-technology-report-2018.

Glickman, T. S. (2000). Glossary of meteorology. *American Meteorological Society*. ISBN 978-1-878220-34-9.

Hall, P. A., & Soskice, D. (Eds.). (2001). *Varieties of capitalism: The institutional foundations of comparative advantage*. Oxford: Oxford University Press.

Held, D. (1995). *Democracy and the global order*. Stanford: Stanford University Press.

Held, D. (2006). *Models of democracy* (3rd ed.). Stanford: Stanford University Press.

Hodgson, G. M. (2015). On defining institutions: Rules versus equilibria. *Journal of Institutional Economics, 11*(3), 497–505.

Jessop, B. (2002). *The future of the capitalist state*. Polity Press.

Jessop, B. (2015). *The state: Past, present, future*. Polity Press: Cambridge and Malden, MA.

Keynes, J. M. (1935). *The general theory of employment, interest and money* (p. 383). London: Macmillan.

Kingsley, D. (2011). Is technology rewiring our soul? *Huffington Post*. Retrieved July 2020. https://www.huffpost.com/entry/technology-spirituality_b_854757.

Korże, B. (2014). Obligations of the social market state and business entities according to the EU guiding principles. *International Journal of Business and Public Administration, 11*(2), 1–22.

Kuhel, B. (2017). *Power vs. influence: Knowing the difference could make or break YOUR company.* Forbes. https://rb.gy/ntjjaf.

Larsen, M. M., Esenaliev, D., Brück, T., & Boehnke, K. (2018). The connection between social cohesion and personality: A multilevel study in the Kyrgyz Republic. https://doi.org/10.1002/ijop.12551.

Latapí Agudelo, M. A., Jóhannsdóttir, L., & Davídsdóttir, B. (2019). A literature review of the history and evolution of corporate social responsibility. *International Journal of Corporate Social Responsibility, 4,* 1. https://doi.org/10.1186/s40991-018-0039-y.

Latter, T. (2017). *How corporate social responsibility drives business performance.* HBR. Retrieved November 2020. https://rb.gy/4r4htg.

Liu, Y., & Ren, D. (2020). China drafts new antitrust guideline to rein in tech giants, wiping US$102 billion from Alibaba, Tencent and Meituan stocks. *South China Morning Post.* Retrieved November 2020. https://rb.gy/vxqrvt.

Luttrell, A., & Sawicki, V. (2020). Attitude strength: Distinguishing predictors versus defining features. *Social and Personality Psychology Compass, 14*(8), e12555.

Mallapaty, S. (2020). How sewage could reveal true scale of coronavirus outbreak. *Nature.* https://www.nature.com/articles/d41586-020-00973-x.

Marr, B. (2020). *Are tech giants with their AIs And algorithms becoming too powerful?* Retrieved September 2020. https://www.linkedin.com/pulse/tech-giants-ais-algorithms-becoming-too-powerful-bernard-marr-1c/.

Martin, E. (2008). *A dictionary of biology* (6th ed., pp. 315–316). Oxford: Oxford University Press.

Marx, K. (1848). Communist manifesto. In Robert Tucker (Ed.), *The Marx-Engels reader* (p. 1979). New York: Norton.

Mazzucato, M. (2018). The value of everything: Making and taking in the global economy. *Public Affairs.* ISBN 978-0-241-34779-9.

McKinsey. (2020, January). *How to ensure artificial intelligence benefits society: A conversation with Stuart Russell and James Manyika.*

Michalos, A. C. (2014). *Encyclopedia of quality of life and well-being research: Social sciences* (Wellbeing & Quality-of-Life). Springer: Dordrecht, Heidelberg, New York, London.

Mirollo, R. E., & Strogatz, S. (1991). Synchronization of pulse-coupled biological oscillators. *SIAM Journal on Applied Mathematics, 50*(6), 1645–1662.

National Scientific Council on the Developing Child. (2020). *Connecting the brain to the rest of the body: Early childhood development and lifelong health are deeply intertwined* (Working Paper No. 15). Retrieved September 2020. www.developingchild.harvard.edu.

Nielsen, R., Akey, J. M., Jakobsson, M., Pritchard, J. K., Tishkoff, S., & Willerslev, E. (2017). Tracing the peopling of the world through genomics. *Nature, 541*(7637), 302–310.

Noorden, V. R. (2020). The ethical questions that haunt facial-recognition research. *Nature.* Retrieved November 2020. https://rb.gy/7ys0ke.

Nowak, M., & Highfield, R (2011). *SuperCooperators: Altruism, evolution, and why we need each other to succeed.* New York, NY: Free Press.

OHCHR. (2011). *Report of the Special Rapporteur on the promotion and protection of the right to freedom of opinion and expression,* Frank La Rue. A/HRC/17/27.

OHCHR. (2020). *Racial discrimination and emerging digital technologies: A human rights analysis.* Report [A/HRC/44/57] - Report of the Special Rapporteur on contemporary forms of racism, racial discrimination, xenophobia and related intolerance. Retrieved November 2020. https://rb.gy/qehwpi.

Offe, C. (1984). *Contradictions of the welfare state* (edited by J. Keane). MIT Press.

Oxford Dictionary. https://www.lexico.com/en/definition/fractal.

Philpot, R., Liebst, L. S., Levine, M., Bernasco, W., Lindegaard, & Rosenkrantz, M. (2019). Would I be helped? Cross-national CCTV footage shows that intervention is the norm in public conflicts. *American Psychologist, 75*(1).

Plotkin, H. (2011). Human nature, cultural diversity and evolutionary theory. *Philosophical Transactions of the Royal Society of London. Series B, Biological Sciences, 366*(1563), 454–463. https://doi.org/10.1098/rstb.2010.0160.

Risdon, C. (2017). Scaling nudges with machine learning. *Behavioral Scientist.*

Rotman, D. (2020). *Why tech didn't save us from covid-19.* MIT. Retrieved March 2021. https://www.technologyreview.com/2020/06/17/1003312/why-tech-didnt-save-us-from-covid-19/.

Saran, S., & Sharma, M. (2019). *Welcome to the age of the platform nation.* World Economic Forum. Retrieved March 2021. https://www.weforum.org/agenda/2019/10/gdp-is-an-outdated-measuring-stick-for-the-new-platform-economy/.

Schiefer, D., & van der Noll, J. (2017). The essentials of social cohesion: A literature review. *Social Indicators Research, 132*(2), 579–603. https://doi.org/10.1007/s11205-016-1314-5.

Schwab, K. (2016). The fourth industrial revolution: What it means, how to respond. *World Economic Forum.*

Schwab, K. (2019). The 4th industrial revolution: What it means, how to respond. *WEF.* Retrieved October 2020. https://rb.gy/hxcllt.

Solon, O., & Siddiqui, S. (2017). Forget Wall Street – Silicon Valley is the new political power in Washington. *The Guardian.* https://rb.gy/vfwzyf.

St. Laurent, A. M. (2008). *Understanding open source and free software licensing* (p. 4). Cambridge: O'Reilly Media. ISBN 9780596553951.

Sun, L., & Yu, R. (2015). The impact of corporate social responsibility on employee performance and cost. *Review of Accounting and Finance, 14*(3), 262–284.

Sustainable Development Goals. (2015). Retrieved October 2020. https://rb.gy/st0o9n.

Tarnoff, B. (2017). *Silicon Valley siphons our data like oil. But the deepest drilling has just begun.* https://rb.gy/ti1ghd.

Therborn, G. (1987). Welfare states and capitalist markets. *Acta Sociologica.*

UN Economic and Social Council (ECOSOC). (2006). UN Economic and Social Council Resolution 2006/23: Strengthening basic principles of Judicial Conduct, 27 July 2006, E/RES/2006/23. Retrieved https://www.refworld.org/docid/46c455ab0.html. Accessed 19 March 2021.

UN Global Compact and OHCHR (Office of the High Commissioner for Human Rights). (2017). *The UN guiding principles on business and Human rights: Relationship to UN Global Compact commitments.* Retrieved July 31, 2017, from www.global-compact.de/sites/default/files/jahr/publik ation/UNGPs_gc_note.pdf.

United Nations Charta. (1945). https://www.un.org/en/sections/un-charter/preamble/index.html.

United Nations Office of the High Commissioner for Human Rights. (2012, June). *The corporate responsibility to respect human rights: An interpretive guide.* HR/PUB/12/02.

Universal Declaration of Human Rights. http://www.un.org/en/universal-dec laration-human-rights/.

Von Bertalanffy, L. (1968). *General system theory: Foundations, development, applications.* New York, NY: George Braziller. Inc.

Von Bertalanffy, L. (1972). The history and status of general systems theory. *Academy of Management Journal, 15*(4), 407–426.

Von Hippel, W. (2018). *The Social leap: The new evolutionary science of who we are, where we come from, and what makes us happy.* New York, NY: HarperCollins.

Wallach, O. (2020). *How big tech makes their billions.* Retrieved November 2020. https://www.visualcapitalist.com/how-big-tech-makes-their-billions-2020/.

Walther, C. (2014). *Le Droit au Service de l'Enfant.* France: Universite de droit, Aix-Marseille UIII.

Walther, C. (2020c). *Connection in times of Covid. Corona's call for change.* Palgrave Macmillan. New York.

Wettstein, F. (2009). *Beyond voluntariness, beyond CSR: Making a case for human rights and justice.* https://doi.org/10.1111/j.1467-8594.2009.00338.x.

World Economic Forum (WEF). (2020a). *From optimism to realism.* Retrieved October 2020. https://rb.gy/pn3vmx.

CHAPTER 4

Exposure

Abstract The status quo is characterized by speed of innovation and lack of compassion. Technology advances ever faster, leaving lives and legislation to trail behind. Beyond pure curiosity and the ambition to expand our expertise, we must gear our undertakings toward the well-being of all people. Everyone has a piece of *Accountability* for the causes and consequences of progress; including inequity and abuse. Since we are all EXPOSED to the situation, we all can contribute to remediate it. Accountability does not relate to the outcome of our efforts, but to the choice of trying. It relates to efforts and the aspiration behind them. Compassionate technology starts with compassionate people. Such technology can then serve to help people become ever more compassionate, kind and generous and to remain that way.

Keywords Exposure · Matrix · Solutions · Algorithms · Human rights · Accountability

Technology is neutral. Its impact depends on the aspirations of those who envision, produce, and use it. Our EXPOSURE to technology is determined by *Accountability* in each dimension of the *mmmm-matrix*:

© The Author(s), under exclusive license to Springer Nature
Switzerland AG 2021
C. C. Walther, *Technology, Social Change and Human Behavior*,
https://doi.org/10.1007/978-3-030-70002-7_4

Micro: Consumers—because their (perceived) desire (in-)directly influences which products (soft, and hardware) are designed and taken to scale.

Meso: Tech companies—because new products create new desires amid customers who adopt and potentially get addicted to these products without intention (or even awareness of their own dependency).

Macro: Governments—because the two aforementioned layers are wrapped into the obligation of the legislator to establish a legal framework that prevents abuse and sets high social standards.

Meta: Supra-national entities—because both, technology itself and the operating scope of corporate tech giants are all-pervasive and borderless. National initiatives are vital, yet they do not solve the issues at stake unless they are globally concerted.

Respectively and together these layers loop back to individuals whose aspirations and actions converge in practice—because individual behavior generates collective dynamics which create social norms. Conversely individual behavior is influenced and fueled by these social norms (Biccheri 2016). Society is an organic kaleidoscope that derives from our behavior; both passive and active. Technology can be a tool to facilitate the ongoing flow of change. *Awareness* of the issues at stake, and their multidimensional repercussions allow us to draw the line between ingenuity and infatuation with our skills, between coherence and complacency, between compassion and comfort. Accepting the context and the available tools is required to tailor a technology that is conducive to society.

Comfort-zones are no home for growth. Settling into that zone means to renounce on a ZENITH before the journey has begun. Travelers do not fail because their goal is too ambitious. Failure to reach the destination of our dreams has two reasons:

1. We do not take the time to clarify our perspective, failing to understand WHO and WHERE we are now, and WHO and WHERE we want to be, which prevents us from identifying WHAT we need to do to get from here to there;
2. We are afraid to identify our WHY.

Both steps are connected, once one is taken the other follows naturally.

When our purpose (Why), personality (Who), position (Where), and pursuit (What) align, the journey of OPTIMIZATION proceeds smoothly, taking us gradually closer to the ZENITH of our personal existence. In that process it contributes to move the communities that we are part of closer to the collective ZENITH of humankind. Progress requires EXPOSURE to the unknown, including the unknown parts of ourselves.

Paradoxically, EXPOSURE to risk is a safe venture if it is undertaken in a PERSPECTIVE of compassion for oneself and others. Because whatever happens along the road serves to refine who we are. Once the road is traced and the journey has begun, technology may serve to keep us on track.

Society can be configured to bring out the best of and for every part of it. But that will not happen by accident, nor should it be a default definition of reality that is designed by a selected few in a sci-fi bubble. Developing the global technological architecture for tomorrow is not a centralized endeavor. The (unfulfilled) potential of technology is locked by the aspiration to meaning of the individuals who shape and use it. The ambitions of meso- and macro-players must be steered by citizens; by visionaries and volunteers, customers and coders who are conscious citizens. The outcomes of technology depend on the objectives toward which the minds that design, deliver and digest it are geared. "The oldest maxim of computer science is Garbage In, Garbage Out"[1] (Fuechsel in Rouse 2008).

As it is, we have 4 options. We can

1. Charge ahead blindly, forgetting about lessons that could have been learned from (failed) promises that did not deliver in the past, such as the Green Revolution that has not prevented millions from starving until this day.[2]
2. Pursue the path of inertia, clinging with complacency to a failing status quo—gradually going down like rats on a sinking ship, like every leading civilization before us.
3. Contemplate the perspective proposed in this book, and appreciate it without action.
4. Give the present perspective, and thus ourselves, a fair chance to prove itself.

Regular check-ins with our aspirational compass offer grounding amid uncertainty; because purpose serves as a NorthStar in a rapidly changing environment. Awareness of humanity's shared foundations including the moral blueprint and the sameness of our multidimensional nature offers stability. In difference to technological progress Generosity, Compassion, Honesty, and Courage are not subjective, transitory phenomena. They have survived the tides of time and will be valid as long as humanity is alive.

In the following sections we first look at two types of technology that may serve at the individual and collective levels, respectively. *Valuable Wearables* address the unfulfilled micro-potential. Seeing the double nature of individuals, as 4-dimensional organisms on the one hand, and as people who occupy a role in society, Valuable Wearables contribute (in)directly to address gaps at the meso-, macro-, and meta-level. Tailored toward the latter three, *Aspirational Algorithms* map and optimize flows within institutions and between them. AA are configured to mainstream quality of life for all, using the latent potential of meso-, macro-, and meta-entities (Fig. 4.1). Following these technological tools, we conclude with a set of suggestions for each part of the *mmmm-matrix*. Neither people nor Planet Earth do need this type of technology to thrive, but both can benefit from it.

4.1   Valuable Wearables. Digital Coaches

Software propels hardware, hardware supports software. Human aspirations, emotions and thoughts impact the evolution of a person's body and behavior; whereas the physical sphere influences how they feel and think. The same logic applies to technology. Using the body (human hardware) as an entry-point, wearables (technological hardware) can nurture behavior patterns and attitudes (human software) depending on the apps and design-intentions (technological software) that go into them.

Depending on their level of sophistication wearables already monitor physiological elements (heartbeats, steps, oxygen absorption, sleep cycles) and mental/emotional tendencies such as stress. As seen earlier, companies have also begun to develop devices that directly tap the brain with the promise to improve performance, fight pain, calm the user and accelerate the process of learning new skills (Swah 2019). This trend can be seized to nurture awareness of the 4 dimensions, and to cultivate consciously a pro-social attitude.

Designed with the intention to foster emotional intelligence, inner peace and compassionate behavior wearable gadgets can serve in different ways. Let us envision a range from 1 (low) to 4 (high) intensity:

> Stage 1—Equipped with an *Emotion Early Warning System* (E-EWS) the device uses heartrate and body temperature to detect when we are stressed. An alert is triggered before we enter the 'fight/flight mode' where rational thinking is switched off (Cannon 1932). Detecting signs of anger, the E-EWS emits a vibration to snap the user out of their acquired behavior pattern; giving them the opportunity to take a couple of breaths and calm down before re-acting hastily.
>
> Stage 2—Conceived with the intention to go against the urge of bias and heuristics, and using technology to monitor impulses in the brain, the E-EWS triggers an alert when the user is about to fall prey to sunk cost fallacy, representation heuristics, the Dunning Kruger effect, etc.[3] The alert is combined with suggestions of alternative perspectives. By sensitizing users to the likely consequences of their action, the brain's "System 1 is prevented from taking over System 2" (Kahneman 2011). The E-EWS may help to enlarge the scope of the most apparent conclusion by highlighting silent evidence and forecasting fallacies.
>
> Stage 3—Programmed based on the users' values and aspirations the E-EWS tracks actions that are in and out of synch with these ambitions, emitting reminders and recommendations in situ to promote the alignment of aspirations (identified and inserted by the user into the system) and behavior. Nudges are released that address the subconscious mind. Priming users toward re-actions that are conform to their own intentions.
>
> Stage 4—Building onto the intra-wearable technology used in COVID-19 contact-tracking apps[4] the E-EWS detects signs of sadness or anguish in other people. When they enter the radar of the user the Wearable alerts the user and suggests ways to address and possibly alleviate the suffering of the other person. Within a reward system similar to video games, bonus points are collected for each action that is geared to enhance another person's wellbeing. Kindness becomes a game that helps and happifies.

Envision a versatile watch that is not just smart but kind; which is not just tracking and encouraging the steps and stairs you take, but the gestures of kindness you make. Like a Fitbit or Apple Watch, where you log your daily food-intake while automatically tracking workouts, etc., the *Valuable Wearable* is designed to keep track of your interpersonal behavior. A real-life diary helps you log your mood, and anything from simple acts of kindness such as a nice word to lift the mood of a co-worker, to more complex efforts such as organizing a fundraiser. It offers a way to reflect and remember what you are grateful for.[5] Self-tracking is completed by biometric tools that automatically measure heartbeat, blood-pressure, hormone emission, etc., to identify correlations between kindness toward others and your own mood and wellbeing (Positive reinforcement increases our motivation to continue a chosen track (Schultz 2015).[6]

Research has shown that our behavior changes when we feel observed (McCambridge et al. 2014). This applies even when the observer and the observed are one and the same. Keeping track by using a metric like counting certain activities is enough to influence how much/little of these tracked activities are undertaken. If the behavior is perceived as valuable (i.e., going to the gym) their frequency increases, if it is perceived as avoidable (i.e., overeating) we make (more) efforts to reduce/avoid it.

Using the so-called Hawthorne effect[7] we can use the body to reach the mind and hereby strengthen emotional willpower in the pursuit of aspirations. Because they are part of the same organism and complement each other. Following the logic of undivided wholeness, "our consciousness derives from the implicate order that underpins the explicate order of the universe as we experience it" (Bohm 1980). Mind and matter are projections into our explicate order from the underlying implicate order. Awareness of this inherent connection leads to genuine EXPOSURE to the reality that is.

4.1.1   Digital Coaches

Machines may support us in the ongoing struggle with the part of ourselves that wants comfort and minimal effort. Serving as a neutral intermediary and 24/7 guide, *Digital Coaches* may smoothen the transition from aspiration to action.

The aspiration to be or at least become a 'good' person is built into our DNA. We evolved to cooperate with those in our tribe and beyond (see Sects. 1.2.4 and 3.1). Yet our day-to-day thinking and feeling may be flawed by material needs and desires. Thus, while it is sometimes challenging to manifest the 4 core values (generosity, compassion, honesty, courage) it is never out of reach. We know (mind) that it is good to exercise and eat healthy, to be kind, honest and generous. Our recurring New Year's Eve resolutions reflect that we want (heart) to reflect this theoretical understanding in our de facto lifestyle (body). And yet, this recurrence also illustrates that we regularly stray from that path that we *know* to be the right one.

Machines are free from the filter of emotion, memory and bias. Equipped with sensors they have begun to detect emotions in humans and to react in an appropriate manner. But they do not generate emotions themselves. Immune to sensorial inputs, they experience neither love nor hate. Their action is free from greed, revenge or jealousy. Unless this feature is specifically built into their algorithm robots are unable to act in dishonesty. Machines make mistakes because of human mistakes in their coding. Depending on their operating system they can make truly neutral decisions in line with the aspirations that underpinned their design. Unlike us, who aspire to something, decide to do it, and then change our mind along the way (Schwartz 2004) machines are configured to stay on track.

Technology can help us do the same; reminding us of the aspects that matter to us when we are clearheaded. Thanks to artificial agility Apps can adjust to the user's evolving environment. Like a coach they may ask questions and possibly give hints to trigger thinking, however the decision to choose this or that course of action remains with the user.

Imagine an App featured like Amazon, yet instead of nudging you to check out this product or that sale, it pushes you to remember the quintessential questions of your existence (Why are you here? Who are you as a person? Where do you stand in your life and where do you want to be? What do you have to change to align your aspirations and actions?) Imagine pop-ups that do not seek to sell you something, but remind you of your aspirations when you are about to snap into counterproductive actions.

Presently our voice-assistants are geared toward commercial goals. Those behind the screen prevent a change of their scope. Alexa, Siri, Google and Cortana could be reconfigured by their inventors to serve, or at least stop impeding, personal growth. Alternatively, a new assistant can be built—a virtual Socrates, a coach in the pocket whose only purpose is to nurture the user's best self, and hereby the environment in which he/she operates.

4.2   Aspirational Algorithms (AA)

"Aspirational Algorithms" (AA) are a type of technology designed to achieve the best possible outcome for all involved parties (Fig. 4.1). They are a technological reflection of humane intentions for humanity. Programmed with the aspiration to distribute resources equitably AA serve to place everyone in the conditions required for quality of life; while minimizing conflict arising from the (re)distribution that is required to establish and maintain that (temporary) optimum. It may support humans by pulling their resources within a neutral framework of thought that is based on common denominators. Thus, countering the challenge of people to collaborate; and to stay on track in the pursuit of an objective.

Society changes and so do we. "The increase of disorder or entropy is what distinguishes the past from the future, giving a direction to time" (Hawkins 1988). Only the basic parameters of our being and becoming remain intact (Chapter 1). They are the same for everyone and offer stable planning components to move forward.

The PERSPECTIVE that we are at core all the same is the point of departure for the required OPTIMIZATION of resources, in the understanding of a shared responsibility for the fulfillment of every person's human needs and rights. Although nobody knows the future, this does not prevent us (1) from establishing clarity concerning the future that we *want*, and, in a reverse-engineering process, (2) to design and implement a scheme that takes us towards such a future. Based on the *mmmm-matrix* technology can map the twice 4-dimensional underpinnings of human life, showing connections and complementarities.

4.2.1   Added Value of AA Versus AI

The use of systemic thinking to shape social solutions is not new; nor did it yield transformation in the past as the present illustrates.[8] The combination of 4 new features makes a difference this time:

- Appetite—Billions of people from all ways of life see, know and feel that the current path is not sustainable. They are ready for change.

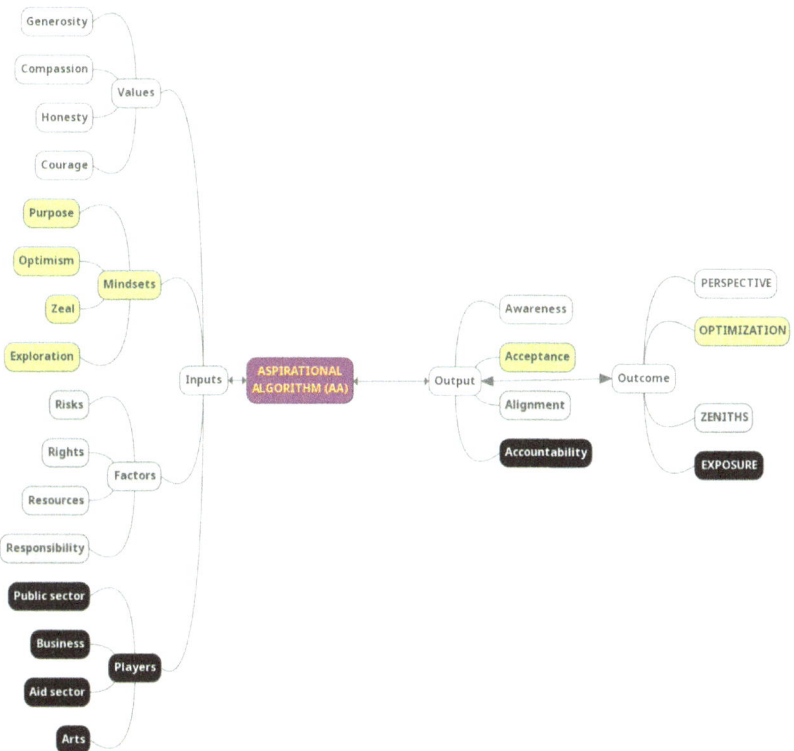

Fig. 4.1 Aspirational Algorithm. Technology is neutral. The deriving Outcomes (differently said its influence and impact) depend on the Values and the Mindset of 4 groups of individuals: visionaries/sponsors, designers, regulators, and users. Conscious citizenship that is anchored in values and propelled by purpose is the name of the AA tech game. If the underpinning orientation is geared toward the best possible results for humanity and Planet Earth, then technology, possibly in the form of *Aspirational Algorithms* results initially in Outputs that are beneficial, and ultimately, based on these Outputs, it generates Outcomes that are sustainable and meaningful along the journey from now to our respective and collective ZENITHs

- Advancement—Technology is more sophisticated than ever before in human history.

Joint with two attitude shifts these circumstances render the present proposal a winner:

- Awareness of the twice 4-dimensional dynamics that underpin society and unite individuals based on the same parameters
- Acceptance of the accountability that derives from it.

If we shape and use technology with the intention of positive consequences for the whole of humanity, the Outcomes follow accordingly. Configured accordingly technology may even compensate for the caveats of human bias during the implementation phase—

In 1738 Dutch polymath Daniel Bernoulli devised an equation to make the perfect choice under conditions of uncertainty. "The expected value of any of our actions – that is, the goodness that we can count on getting – is the product of two simple things: the odds that this action will allow us to gain something, and the value of that gain to us" (Gilbert 2005).[9] And that's where the problems start. Because our inner valuation parameters do not remain the same over time. What seemed better before, appears as worse from an ex post perspective. Biases comes into play, such as references to facts that are the most available in our mind (but which are not necessarily the most relevant to the question) which render certain (desired or despised) outcomes more likely than others, etc. As seen earlier, human beings are not primordially rational decision-makers; we are not equipped to maximize our choice potential, because our abilities are limited, and biased when it comes to absorbing information about options, and then ranking these options from least to most beneficial.[10]

"The human mind assembles a few relationships to fit the context of a discussion. As the subject shifts so does the model. When only a single topic is being discussed, each participant in a conversation employs a different mental model to interpret the subject. Fundamental assumptions differ but are never brought into the open. Goals are different and are left unstated. It is little wonder that compromise takes so long. And it is not surprising that consensus leads to laws and programs that fail in their objectives, or produce new difficulties greater than those that have been relieved" (Forrester 1971). Technology is neutral and focused; humans are not. The more complex the issue at stake, and the more uncertain the context in which it takes place, the more likely it is that decision heuristics come into play (Heiner 1983). And then there is inertia. The "yeah, whatever" heuristic (Thaler and Sunstein 2008) relates to the human tendency of going with the default option, not because it is the best, but the easiest. Complacency strikes.

Technology might offer a way out of the dilemma posed by unlimited unsatisfied optionality, default derailing and straight bad choices. Standard Artificial intelligence (AI) like the one currently used can address the constraints that derive from the limited human capacity to solve a difficult problem. Thus, it may promote enhanced, more participatory, and effective policy and programming. Framed accordingly, it can generate insights regarding the root causes of problems and the likely effects of possible solutions. But an important implication of 'bounded rationality', mainly in evolutionary economics, is that there is a second constraint. Even if humans had full rationality, it would not be possible to optimize outcomes because the objective itself is a moving target (Simon 1976).[11] There is no one single ZENITH. A different kind of algorithm, based on the present paradigm, is therefore needed to help humans deal with the second issue, because moral and aspirational factors are part of the equation.

Whereas AI is neutral at best and tailored at worst to serve commercial interests, Aspirational Algorithms (AA) are designed to fit 4 types of choice-constellations, whose aspirational intensity increases from 1 to 4:

> Type 1—At the lowest intensity level AA offer a large range of options for users to choose from, based on their own personal, previously observed choices. The AI for this option is already in use, leading for example to the type of ads that pop up based on browsing history, etc. The only differences of AA at this level (type 1) versus current AI are its orientation (altruistic rather than commercial) and focus (multidimensional and intra-connected instead of biased to certain providers).
>
> Type 2—At the next level AA makes choices based on options selected by users, or based on a set of pre-defined criteria. It frees human decision-makers from the dilemma of choosing one option among others for questions that are recurring and are prone to

be addressed by a mix of inertia, emotions and present perspective bias, so-called "routine decisions" (Nelson and Winter 1982). The advantage is the technological possibility to make objective and mathematically meticulous comparisons between different options, to make the best choice—without succumbing to bias, emotional rerouting and lower-lying motivations such as greed and jealousy.

Type 3—A hybrid of Type 1 and 2 allows to delegate the selection (filtering) of options to choose from, and the choice itself to AA, based on objective criteria and values as pre-determined by users. Unlike currently used AI, it is not based on previously observed choices (i.e., the individual is passive) but coded to manifest values that have been explicitly recognized and stated by the individual. Users have an active role in shaping the algorithm along the values that matter to them.

Type 4—The optimum use of AA may be its potential to not only compile and chose from the existing range of options; but to design solutions that had not and possibly could not have been envisioned by the human mind (due to the constantly evolving environment and the gigantic data quantities involved in monitoring and mapping the consequences that derive from that evolution). A feedback-loop from AA to the user leaves the decision of last resort to the human. Designed to systematically cultivate progress toward Augmented Humanity (AH), AA type 4 seeks to systematically eliminate the gap between the status quo and its optimized counter-face. AA may serve in complementarity to current forms of AI, by dealing with issues that entail values and redistribution in situations where no objective solution exits that everybody would agree with.[12]

Technology makes it possible to combine resources beyond material and immaterial boundaries. It depends on humans if that happens, and whether the impact is beneficial or disastrous. "AI is like a mirror, and if you don't like what you see in the mirror, of course, you can start to look for some distortions, or you can look at yourself and start changing your behavior. [...]" (Kasparov in Miller 2020).

****

Human Aspirations (HA) that reflect a holistic PERSPECTIVE and are geared toward the OPTIMIZATION of human potential eventually result in Augmented Humanity (AH). Aspirational Algorithms (AA) are designed with this mindset. They do not replace human effort but facilitate progress toward the state that appears as the ZENITH based on current knowledge. Whether dystopia, or equitable expansion, the impact of technology depends on coders and consumers; both are part of many communities, and ultimately citizens of Planet Earth. It is not about quantity—more tech, more bytes, more x. The required shift is qualitative, including changes in individuals behavior, access to goods/services, and the circulation of profits. The sequence starts with Human Aspirations (HA), moves via neutral Artificial Intelligence (AI) toward pro-social Aspirational Algorithms (AA); which help us on the road to Augmented Humanity (AH) (Chapter 3, Fig. 3.1)

4.2.2   AA Prototype

Technology can go in many directions. The challenge is to design a system that

1. Asks the right questions;
2. Deciphers the resulting information to come up with ambitious, creative yet pragmatic answers;
3. Validates the resulting outputs based on human values; and
4. Facilitates the process from ideation via planning to implementation and monitoring.

The following software outline is designed to systematically nurture the best version of human users. It is an overview of the inputs and expected outcomes of such a tool. Benefitting from the logic of Deep learning this type of AA evolves with the human users; yet while rising with them it prevents them from regressing. As a result, the capacity of both to care and share grows to grow further.

Input—Building an AA requires 4 building blocks, which may be expanded based on local context:

- *Values*, starting with the 4 core values described earlier—Generosity, Compassion, Honesty, Courage.
- *Mindset*, including 4 basic mental parameters: Purpose orientation that transcends personal interest, Optimism for the future, Zeal to

achieve the best outcomes for humanity, and Exploration of opportunities with the ambition to move beyond forgone conclusions. These relate to the overall aim of OPTIMIZATION yet go deeper by establishing a more granular mental matrix.

- *Circumstances*, including Risks, Rights, Resources, and Responsibilities; covering material and immaterial factors.
- *Players*, the main stakeholders whose attitudes and actions influence the outcomes—Public sector, including Academia; Private sector; Aid organizations; and Artists, including Media.

Big data may serve for the profiling of *Circumstances* and of the *Players* involved by using its signature strength—the four Vs (volume, variety, velocity, veracity; see Chapter 3). Herein the added value is Big Data's ability to map and highlight connections; which may serve to overcome a different V—the Vanity of researchers who consider that a reality that opposes their hypothesis is mistaken. Epistemic humility is an essential ingredient of radical creativity (Matthew 2006).

The results of AA are short, medium, long-term, and ongoing:

Short-term Results include the immediate influence on users. Configured along the logic of the *Scale of influence* (Sect. 1.4.1) and addressing the 4 individual dimensions of being alive, users targeted by AA end up:

- Inspired, because their personal aspirations are titillated and awoken.
- Induced emotionally, thus stirred to shift from understanding issues to caring about them.
- Intrigued intellectually, hence pushed to question the status quo and contemplate alternatives.
- Ignited to get proactively involved, either physically offline or online.

Medium-term Outputs deriving from the first impact stage are a transition to the final outcomes yet have their own value. The AA helps users evolve within their own 4 dimensions. It is furthering their:

- *Awareness*, reminding them of the dimensions that influence their experiences and expressions;

- *Acceptance*, illustrating the links between these dimensions and the unescapable nature of the 4 principles (connection, change, continuum, complementarity. Sect. 1.1).
- *Alignment*, nurturing connections—inside out and outside in, through positive reinforcement.
- *Accountability*, making them grasp the (potential) consequences of their choices by visualizing the impact of their (in)action.

Long-term Outcomes build on the preceding two stages, with lasting shifts of user attitude and action. AA nurture the transition, by

Stage 1—Showing in real time the connection of causes and consequences in the interplay of individual and collective dimensions; hereby the user's PERSPECTIVE is sharpened and expanded. *Awareness* is cultivated.

Stage 2—Pointing out the best possible choice based on the criterion of minimum harm and maximum benefit for the user, others, and society; while nudging users to follow a path of OPTIMIZATION by making it as easy as possible to follow that course of action. *Acceptance* is built.

Stage 3—Encouraging users to not only acknowledge, analyze, and accept the mutual interplay between themselves and others, but to address it; hereby encouraging them to stay on course to their personal ZENITH. *Alignment* is nurtured.

Stage 4—Proposing a range of unexplored options to maximize the benefits for themselves and others, hereby persuading users to risk EXPOSURE to uncertainty, and venture beyond their comfortzone. *Accountability* becomes tangible.

****

There is a human alternative to AA optimization.

Based on *Awareness* of the multiple dimensions of our being, and of the principles that operate underneath them, we can choose *Acceptance*, and based on it, seek the systematic *Alignment* of the dimensions that influence us, by influencing them. When our action is not a default derivate of circumstances but a deliberate choice, we shift from passive to active players in the game of life.

We cannot make unknown what has entered our mental sphere. Awareness of the PERSPECTIVE that is presented in this book offers the choice to (not) use this awareness and the influence that derives from it. In this context, *Accountability* does not relate to the outcomes of influence—which are never entirely in our hands, but to that choice itself. When we have the means to meet a certain end and do not deploy them, we are accountable for this omission and the consequences that derive from it.

Technology will not save us. It can support solutions that are initiated by humankind. Neither more nor less.

To be beneficial to humankind both, technology and the regulations to frame it, must reflect the aspiration to serve the common good and the values that underpin it.

4.3   From Potential to Personal Power

Not everything that can be imagined should lead to execution; but once the intention is compassionate the outcome is likely to reflect the values of the one who acts, while addressing (the perceived) best interest of the one for whom the action is taken for. A society that is worth living in hinges on ideas that are planted with seeds of aspiration, rather than pecuniary ambition, in a soil of human values rather than a heap of banknotes. Technological propensity and human imagination must operate within a grid of universal human values. The manifestation of these values mattered yesterday and will tomorrow. Progress does not render them obsolete, otherwise it has defeated itself.

Ironically, our ever-increasing possibilities to connect with people and information leave us ever more disconnected from each other, and from reality. We communicate more and say less, talk a lot and feel but little.[13] Gradually we are getting attuned to observe, being observed, and observing the observation, turned into bystanders of our own life. Rather than EXPOSING ourselves to reality, we are stepping away from it. Spending our time liking and sharing online, we are increasingly absent offline; missing in situations where our generosity and compassion, our honesty and courage would make a difference for someone. Simultaneously our ability to concentrate on new information, to analyze and assimilate it decreases due to information overload (Carr 2010). We are settling into the echo-chamber of our networks, confirmed in our view of the world; protected by the membrane of our online comfort-zone we are protected from offline facts, reality. Taking the leap from complacency to creativity does not require more technology, but a different attitude toward the one that exists.

As seen earlier we are prone to succumb to the inbuilt tweaks of our mental system, from fallacies and bias to emotional hijacking; from complacency to default choices (Chapter 1). Using that candid PERSPECTIVE of our own operating model we look in this section at ways to not only compensate but harness the humanness of our own system with one stated intention—Augmented Humanity. Addressing the 4 dimensions of society the proposed initiatives are complementary in their contribution to a mindset of generosity, compassion, honesty, and creativity.

Technology can nurture and amplify a virtuous cycle, linking human aspirations and aspirational algorithms, but it cannot initiate that cycle. The process of optimization may include Valuable Wearables, Digital Coaches, and *Aspirational Algorithms*—or eschew technology altogether. Only humans who are purpose-oriented and value-based can create a type of technology that is conducive to Augmented Humanity, thus the following recommendations for each one of the *mmmm*-dimensions zoom in on human mindsets. The list below is not exclusive but a selection of options to get started.

4.3.1   Micro-Dimension

4.3.1.1   Mind Muscles

Thoughts influence emotions which impact behavior. Thus, training the mind is an investment that pays off in terms of our own quality of life, and that of others; because it impacts how we perceive and address our environment. Mental change does not hinge on high-tech. Solutions inside our natural set-up exist, at no cost. Meditation "means familiarization with a new way of being, a new way of perceiving things, which is more in adequation with reality, with interdependence, with the stream and continuous transformation, which our being and our consciousness is" (Ricard 2004). We are the masters of our mind and thus of our future. It would be sad to relinquish this power to machines; even if they were invisible implants that make us mindful cyborgs.

4.3.1.2   Connection

Quantitatively speaking we are ever more connected; qualitatively we are moving apart, from ourselves and others. One challenge is our failure to cultivate comfort with solitude. "Solitude is where you find yourself so that you can reach out to other people and form real attachments. When we don't have the capacity for solitude, we turn to other people in order to feel less anxious or in order to feel alive. When this happens, we're not able to appreciate who they are. It's as though we're using them as spare parts to support our fragile sense of self" (Turkle 2012). We like to believe that constant connectivity overrides our loneliness; that more channels make us less vulnerable. But the opposite is true. Unable to be alone, we are always lonely. We can consciously train our ability to listen and hear, to communicate and 'connect' from the inside. These skills must be taught to children from an early age.

4.3.1.3   Perspective Polishing

If an individual has been raised in an unilateral understanding of the world, it takes time to form a new perspective. But it is possible to systematically retrain our awareness of reality. A holistic lens changes the understanding that we have of our Self and others. Recasting our role within a constantly changing continuum allows us to not only understand intellectually the added value of The Golden rule of reciprocity, but to genuinely desire its application. Retraining our perception is a conscious effort (Sect. 1.4.3). Which has benefits. Helping others promotes physi-ological changes in the brain linked with happiness (Otake et al. 2006), while enhancing our immune system and ability to withstand stress (Post 2014). In addition, it improves our support networks and self-esteem (Pressman et al. 2015). Doing good is good for us. The trick is to get started.

4.3.2   Meso-Dimension

4.3.2.1   Inspiration Islands

Keeping ideas alive is easier with a physical anchor. This is the reason for churches and temples, mosques and shrines. For centuries libraries have served as safe havens for readers and thinkers. Whether dedicated to knowledge or spirituality, the physical space does not contain the meaning. It is merely a vessel; to nurture meaning among those who visit it. We are social beings who thrive in the presence of others. Even introverts need a balance of alone and togetherness (Hills and Argyle 2001).

Connecting the like-minded can happen offline and online; ideally both realms nurture each other. Inspiration Islands may help to bridge the divide between people, but also between aspiration and action; connecting meaning, mind, and the materialization of ideas about change into proactive transformational steps. In times of online platforms, physical libraries are at risk of extinction. They may be repurposed as Inspiration Islands that offer a secular haven for seekers, thinkers, and doers who want to make a difference but are unsure how to break out of their isolation.

4.3.2.2 Compassion for Change

Organizations that inspire the public are constituted by inspiring people. Staff are influenced by the priorities, positions, and programs pursued by their employer. The *P-Puzzle* shapes institutions (Sect. 3.2). The often-stated belief that systems prevent people from living according to their values in everyday work life is a tautological argument. 'Positive engagement' establishes a parallel between the meaning that employees see and pursue at work, the intensity of their engagement, and their overall happiness (Stairs and Galpin 2010). The wellbeing of staff conditions workplace happiness, and vice versa. Both are conducive to productivity and performance. Most adults spend a significant amount of their lifetime working. How they feel at and about that work influences their quality of life, and the performance of their employer. Lack of (com)passion at the workplace puts people at risk of burnout and a decrease in deliverables (Kompanje 2018). When people navigate a void of meaning their wellbeing declines, and with it the outcomes of their employer. This matters in every sector, yet in particular within organisations that are dedicated to help others. The latter includes supranational institutions such as the UN, that were established to make the world a better place.

A bridge to connect the individual quest for purpose with the organizational mission can turn a vicious dynamic into a virtuous cycle that feeds off itself. A methodology to induce institutional change via personal transformation is the *Compassion for Change (C4C)* approach. Though applicable in any institutional setting its focus is on entities that have the vocation to promote positive social change. In contrast to exclusive leadership approaches it addresses every staff, with the understanding that internal renewal is conditioned by active top-down and bottom-up involvement. C4C is tailored to reconnect staff with their personal aspirations, and to connect them through these aspirations with the mission of their employer Drawing on the multidisciplinary POZE framework it combines behavioral insights, human psychology, local culture, company values, and lightweight pragmatic tools. In opposition to generalized blueprint approaches C4C invests in local resources to facilitate ownership and low-cost expansion (Walther 2020b).

Using a blended approach of virtual and face-to-face interactions *Valuable Wearables* are not required yet may serve to take C4C to scale at low cost, while supporting sustainable results. The availability of 24/7 (digital) coaching supports each staff during the process of personal change and institutional transformation. *Aspirational Algorithms* can be useful to map staff interactions. Zooming in on sensitive bottlenecks that require change they thus allow to systematize the process at local and larger levels.

4.3.3   Macro-Dimension

4.3.3.1   Socratic Schooling—Reshaping the Education System

Education can increase the quality of a person's life—if it connects learning about life with living (Dewey 1938). Taken that way it can open the door for the individual to dare, dream, and do things that influence society in a positive way. If education addresses the soul, heart, mind, and body it enables young people to actively participate in their community, which is central to subjective wellbeing and collective prosperity. Meaning making must begin early, when the mind is the most malleable. Within the *mmmm-matrix* schools are part of the meso dimension. However, to mainstream the orientation of schools toward a 360-degree understanding of education and experience, national curricula must be tailored accordingly. That is a task for the State. The outline of teaching and learning must guide children to think critically about the mutual influence of their aspirations, emotions, thoughts, and sensations; and about ways to optimize these interplays intentionally.

4.3.3.2   Purpose for Power

In many countries, particularly in poorer areas, education is based on a traditional top-down model, which primarily seeks to inculcate passive knowledge. Performance, not purpose, is the central pursuit. One way to challenge that status quo is the *Purpose for Power (P4P)* approach. It nurtures a mindset that questions context, identifies patterns, and accepts connections. Based on neuroscience and psychology, free from religion and dogma, it can be applied everywhere, using local resources and expertise. Tools address the 4 dimensions of the human being in a holistic manner that is age appropriate. P4P facilitates the shift from a passive attitude toward one's own life to a proactive stance in society (Walther 2020a). Serving as an update of past practices P4P represents a harmonic addition to long standing teaching traditions such as those purported by Steiner, Emilia, Montessori, Dewey, and Piaget. All of which aim for the combination of understanding and experience, to equip children for the pursuit of purpose in life. Setting them on track to their best self these methodologies and P4P place the focus on dignified autonomy. This serves individuals and society alike, by unlocking the individual's inherent potential.

A blended approach, using massive open-online courses (MOOC) in combination with local volunteer pools, could make this type of learning accessible around the world at no or very low cost. Combined with *Valuable Wearables* to guide teachers and students on an ongoing basis, a coach in the pocket may serve to render the results sustainable over time. Another cost-efficient way is an integration of the P4P logic in existing school curricula. This is particularly relevant for countries where large parts of the population remain without Internet access. As emphasized for C4C, technology is neither needed to start nor to scale P4P—it may support it once the will to begin is established.

4.3.3.3 Generalized Generosity

A Universal Basic Income (UBI) addresses human needs with dignity. As most of the recommendations in this list, the idea of an UBI is not new, it has been around since the eighteenth century. It refers, in its purest form, to an unconditional government guarantee whereby each citizen receives a minimum income. The intention behind this payment is to cover the basic cost of living and, thus, to provide financial security. To counter concerns regarding the destruction of jobs through automatization, various Silicon Valley leaders embraced the UBI logic from 2018, either as a universal approach or for target population groups (Ghaffary 2019). Funding examples include the increased taxation of high incomes (Wright 2016) and additional taxes on products/services that are luxury items, including topnotch technological devices, or on financial stock transactions

By early 2020, small-scale localized tests had been run in China, India, Namibia, Iran, Kenya, The Netherlands, Germany, Sweden, Finland,

Brazil, Canada, and the US. Two countries, Mongolia and the Islamic Republic of Iran, did run a national UBI for a short period of time. The results conclusively contradict concerns that an UBI would disincentivize individuals from working and deprive them of the meaning that derives from paid work, while cheating their country out of productivity (Samuel 2020). The results also discredit the notion that an UBI is unaffordable for countries.

Aspirational Algorithms may serve the dynamic of (re-) distributing resources by identifying needs, resources, and the gap in between while coming up with pragmatic suggestions that take a 360-degree perspective of the challenge. At the same time *Valuable Wearables* can coach politicians and business leaders involved in translating the AA plans into practice, by fostering their willingness to follow through on promises.

4.3.4   Meta-Dimension

What goes in comes out, locally and globally. *Glocal* solutions are designed with a compassionate attitude for an interconnected world. Pro-social human minds combined with technological assets that are designed and operated with a generous, compassionate, honest and couragous attitude cultivate pro-social outcomes.

Our future will be marked by equitable economic growth; or by an implosion caused by crossed environmental thresholds, as the creative process runs amok. The impact of technology on people and Planet depends on Human Aspirations. When a critical mass of individuals accepts their accountability for (not) identifying and thus for (not) using their own potential to the benefit of others, and acts accordingly, their combined influence emerges. Action with the aspiration of the common good creates ripple effects. Because those who witness the shift in behavior are inspired to follow that example (Sect. 1.4). Unfortunately the same ripple effect applies to behavior that is driven by greed.

Technology learns and delivers based on human inputs. That link between input and outcome can be detrimental. Misguided intentions (input) lead to misdirected technology (output) which affect the meta level—Society (outcome). An illustration of the connection in/out is the impact of biased data used to train facial recognition algorithms to detect suspicious individuals (Chapter 3, Sect. 3.1)—flawed input (biased data)/flawed output (wrong alert).

The needed shift is not primarily related to new algorithms, methodologies, and datasets. We need to upgrade the aspirations and values that underpin the design and delivery of these tools. Technology may nurture the seeds of humanity; it cannot (and should not) plant them. Expecting machines to ultimately acquire skills that their human creators did not possess is realistic. To presume that these highly skilled machines will aspire to values that their creators lacked, and thus failed to build into their creations, is not.

The 8 billion people on this planet are widely different and have differing views of what an 'ideal society' looks like. We may neither agree that the 'common good' is to be valued, nor even have the same definition of a 'common good'. Yet do we share a moral blueprint (Chapter 1). Though the fine print may diverge, common elements emerge that must be reflected in technology as we go on. Technology along the lines presented in this chapter may help human progress at the micro-, meso-, macro- and hereby the meta-level. *Valuable Wearables* address the individual user, and *Aspirational Algorithms* look at the meso- and macro-dimension. Together they promote an organic optimization of the meta sphere within a holistic approach to the global *mmmm-matrix*, if human willingness is set.

Augmented Humanity (AH) happens off the grid but is expanded by it. Human Aspirations (HA) influence Artificial Intelligence (AI), which serves in the design of Aspirational Algorithms (AA). However, ultimately Augmented Humanity (AH) depends on Human Aspiration (HA). It may be nurtured by AA but does not depend on it (Chapter 3, Fig. 3.1).

4.4 CONCLUSION—MANIFESTING AUGMENTED HUMANITY

As stated throughout this book—Technology is a neutral force that reflects human aspirations. Problems derive from it when these aspirations are not anchored in core values. Technology expands human influence. The dynamics that we start or do not stop now influence future generations, of humans and of machines. The following invitations to dream may serve as forecasts; their realization depends on us:

Envision—

a wave of technology that is planned, designed, and offered without commercial interest. An open treasure chest that is characterized by the best of all worlds: expertise and assets; multidisciplinary diversity and multicultural understanding; neutrality and humanitarian orientation. Uniting private sector players, academia, athletes, artists, the media, United Nations and government officials, non-profit staff, and unaffiliated individuals of all walks of life, behind the shared objective of inclusive growth, social progress, and sustainable development. Imagine a society where people use their potential to lift others reach theirs. A systematic merger of sectors and disciplines resulting in the combination of material and immaterial excellence. All of it geared toward covering everybody's basic needs (i.e., food, water, shelter, safety, education, health, information, transportation, participation); while nurturing a culture imbued with generosity and compassion, honesty, and courage.

Imagine a society that evolves toward meaning, with early childhood programs and school curricula that coach children from the start onto a journey that is their own. An education system that shapes critical mindsets and creativity, paired with a holistic perspective of humanity. A curriculum that cultivates the need of WHY rather than silencing it. Imagine children growing up in an environment where equality and kindness are openly valued; not as lip-service, but pragmatically day by day. A place where mutual help is the norm, and planetary health protected.

Truth be told, Society does not need technology to achieve that. However, if it is designed with pro-social intentions technology may help to expand the scope, scale, and sustainability of such intentions.

Envision—

a technology that transforms our devices into guides and coaches. Free from commercial interests these are coded with genuinely altruistic intentions. Imagine hardware and software that are designed to make people happier, independently of their socio-economic status, their ability to buy a subscription or the next upgrade. An offer that is put into the public space to reach everyone; not to draw users toward one brand, party, cause or another, but to plant and cultivate desire for meaning.

Conceive of an app that reminds people of the ongoing interplay between their aspirations, emotions, thoughts, and sensations, helping them optimize this ongoing interaction between mind and matter, between themselves and others.

Think of devices that nurture the shift from dependency to freedom, reminding individuals and communities that matter fades whereas meaning lasts.

Envision—

a combination of hardware and software that guides people to activate their personal power of choice, nudging them to take the responsibility and choose how they live life, in the present moment.

Envision—

the use of Virtual Reality to expose users to a context where they experience a 'reality' where people live in perfect harmony; helping them thus experience how this type of reality feels and looks. The combination of Augmented and Virtual Reality may place them in an environment that looks familiar but feels radically different, due to social dynamics that operate based on human values. Helping them believe that this is possible, by making them see what it does (could) look like. Extended Reality may then serve users as a safe playground to experiment with a different, kinder version of themselves.

****

The Future is about Choices

Humans feel compassion for others and passion for a cause. They have the choice to be honest or lie, to be generous or egoistic, to act courageously or cowardly. No other person and no technology can relieve us of the effort that precedes choices and derives from them. The discomfort of change is ours to overcome.

Purposeful change is inherent to Nature. It distinguishes living organisms from machines. Consciousness enables humans to change the course they are on. It distinguishes humans from animals and from machines. Machines follow their inbuilt purpose within the limitations of their operating system. They either perpetuate a status quo in which billions are marginalized, or nurture humanity's course to a ZENTIH.

Augmented Humanity derives from genuine human connection. From a **P**erspective of multidimensional interdependency, we can strive for an **O**ptimization of the existing dynamics and the creation of new ones that gradually lift individuals and hereby society to the **Z**enith of their respective and collective existence. **E**xposure to reality as it is, combined with the experience of generosity, compassion, creativity, and courage is the foundation of a tomorrow that is worth living. 4 spheres of influence arise when we put the concepts that were discussed in this book within one framework (Fig. 4.2). It is up to each of us how we shape and navigate these spheres. Our WHY (purpose) influences WHO (personality) we are and become and WHERE (position) we end up being, individually and collectively. Our perspective is not determined by coincidences,

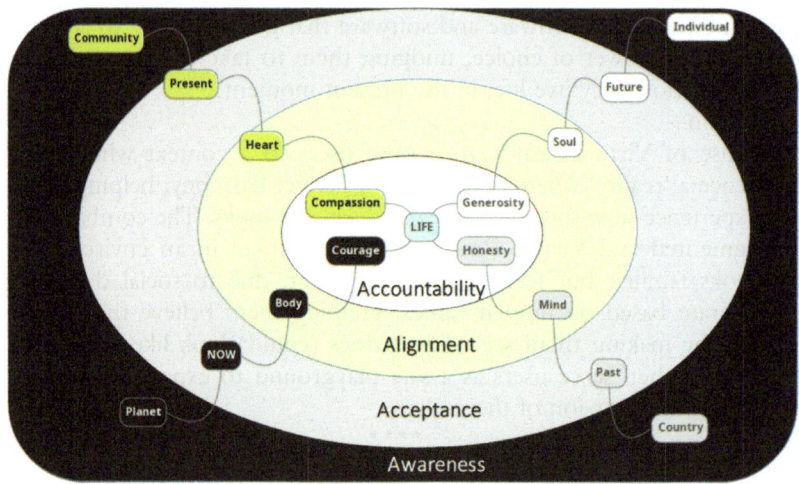

Fig. 4.2 Spheres of Influence. Human existence navigates 4 spheres of influence (Soul, Heart, Mind, Body) which relate to the 4 core values of humanity Generosity (Aspiration), Compassion (Emotion), Honesty (Thought), (Courage) Sensation/Behavior. We envision the future, feel in the present, remember the past and influence the NOW. The intensity of these spheres increases from the periphery to the center. Introduced to a new PERSPECTIVE of life Individuals move from *Awareness* of the dimensions that influence them, via *Acceptance* of these multidimensional interplays to the ability of *Alignment*, where they can consciously OPTIMIZE the dimensions that shape them. Once that level is reached, they can not only systematically influence their own behavior, but also that of others. As their behavior increasingly reflects the synchronization of their aspirations, their emotions and thoughts, their interactions with others become more harmonious. Translating their values in practice they gradually come closer to the best version of themselves, their own ZENITH. Being part of a community, they contribute to that community's progress toward a constellation that lifts individuals to fulfill their potential. As the number of people who progress along these lines increases, Society approaches the collective ZENITH. Individual influence comes with EXPOSURE to reality as it is, and with it the *Accountability* to use one's personal influence for the common good

but choices. WHAT we do (pro-active stance) to align our aspirations and actions influences what will happen. Everything is connected.

The question is not if we want technology to be part of our life, but which role we attribute to it. When we synchronize our aspirational compass and the coding of our technological creations, grounding both in compassion for change that serves the common good, the outcomes will be beneficial.

****

This book concludes with two letters that might be written in 100 years. Which one it will be depends on humans, not on technology.

DIALOGUES 2121

Futurist Letter 1—Banked Backlashes (-)

My Dearest,

I am writing because I will not see you grow up. When you take in these words I am no longer in this realm. The advances of medicine expanded our lifespans ever further, but I have decided that my time has come. I do not want to continue this scam of a life; undergoing one upgrade after the other to this body of mine that has passed its deadline. Science replaced the natural transition from life to death as a result of age, with the whims and wishes of individuals. Clinging on to the illusion of immortality we thus go on, and on, and on. But I won't.

I was born in 2010, part of the last generation that learned to read and write at school. When this missive reaches you, you will absorb it via infofusion. The content will be transferred to your cerebrum upon your 14th anniversary, as part of the yearly mental upgrade that has come to replace primary and secondary education in the late 2060s. It will not be the same as if we met face-to-face, but it is the best I can do.

I cannot go on, but you must know what was. And since deep fake has blurred the lines between true human accounts of the past, and computer-generated fiction, I want you to have one witness account that you can trust.

I was born at a time of momentum. 2021 was a year of sorrow and excitement. The world was going through the aftermath of COVID-19, a global pandemic that infected a billion people before it was vanquished. The so-called 4th industrial revolution, with its fusion of physical, digital, and biological arenas was in full spin. Accelerated by the sheer need of people for safe alternatives to physical interaction during the Pandemic, technology leapt to the center of our life. Shopping and entertainment, communication and education, working and dating, everything moved to the online space.

You never knew another world but be aware that our interactions have not always been through devices. When I was young, we used to meet people in person; we went to schools and offices, visited shops, museums, parks, and cinemas. The transition to a remote reality occurred gradual with an ever-quicker pace from 2021, until offline was the exception and online the norm.

One consequence of this trend is isolation. Your generation has grown into a society that is always on, and never entirely with. This is not entirely new, Already my parents told me about their multitasking exploits— managing phone calls, emails, dealing with me, preparing dinner, etc. But at least they still had the memory of the time before. When there was a time for work and one for leisure, one for communication, and one for silence. During my childhood our devices became ever smaller, until they reached the stage that you are used to. Now everything is 'integrated'. You think about calling and the chip in your head makes the connection to the chip of the person you want to reach; your fridge records shortage, checks the balance of your bank-account and places an order to replace what's missing; the chip in your brain signals boredom and your phone features a range of entertaining advertisements, etc., etc.— you have grown up with this. You know the drill. The problem is that you do not know anything else!

Already I was unable to remember basic math, foreign languages, even phone-numbers of family (yes at the time not everyone's contact was available online; we had private numbers and email addresses)—because we delegated memorization to our phone or computer. But at least I was still going to school. True, since COVID-19 most of this learning happened online (my parents used to tell me about classrooms and games played outside during breaks), but at least it entailed actual efforts on the pupil's part. This changed when brain-chips were introduced. Our inbuilt technological enhancements minimize mental efforts. This leaves a trace in our natural hardware. The damage is not irreversible yet, but soon it will be. Brain scans show the drastic decline of the number and agility of our brain cells compared to previous generations. We allow our brains to gradually atrophy.

At the beginning the shift from offline to online life was perceived as slow, subtle; then it became too drastic. Many researchers raised the alarm bell in the early 2020s. But slowly the cyborgization trend moved ahead. At the beginning those who decided to get tech enhancements did so to overcome a physical handicap; paralyzed people were able to walk again, blind people could see, deaf people could hear, etc. It was amazing. Then increasingly people got upgrades to address other afflictions, like dementia, depression, dyslexia, insomnia, etc. Gradually the qualification of an 'unbearable affliction' shifted. The third wave was that of performance upgrades. Users got brain-chips to expand their 'mental hardware'; no longer was the neurological setting a barrier to a person's skills and stamina. You should be aware that these upgrades entail two parts: first comes the general refurbishment of mental wiring which prepares our operating system for subsequent knowledge transfers and periodic upgrades. Whether it is playing the piano, mastering karate, understanding algebra, or speaking a new language (which is about to become obsolete since simultaneous translation happens automatically through the implant when people communicate in different languages) you pay the chip, you get the skill. Instead of addressing conditions like loneliness with psychotherapy and medication a simulation of brainwaves serves to switch off an undesired state of mind.

First there were few who went for implants; like traveling to Mars which was once reserved to eccentric billionaires and is now a luxury trip like safaris or cruises once were. Then gradually, like cellphones in the 1990s, brain-chips went mainstream. Yet when they became more widely available, implants remained for a while a luxury gadget. Like face-lifting was sought by women keen to eliminate 'wrinkles' (Check online 'natural aging process'. Harvard University is long gone because universities are no longer viable, but their website is still maintained by some nostalgic nerds in their 100s). As more and more people got these mental implants the pressure on those who had resisted grew stronger. The implanters outperformed the rest. Over the subsequent decade the sophistication of these tools and our dependency on them grew exponentially.

Our biological brain is an intricate feature that is geared to constantly adjust and advance; but such organic progress entails personal effort. We do not consciously choose to *not* engage in mental effort, but de facto we do refrain from it; following the path of least resistance we do no longer challenge our innate resources. Consequently over the past century machines have progressed while humanity regresses.

When I was a child the POZE paradigm was all the rage—but it remained a niche perspective. Stating that every individual has a purpose and a unique combination of resources to fulfill it, POZE stressed that individuals can systematically nurture the interplay between their aspirations and actions, between people and planet. It proposed practical steps to do so, and I guess if more people had chosen that track then, I would not quit now. Unfortunately, the majority went for quick fixes; mental upgrades on sale. To my shame I am no exception. Only in hindsight do I know that I, we, missed an opportunity to break the mold of victimhood and dependency.

I am sharing this not to make you regret what is lost. But to show you what was, and still is, possible. I hope it helps you question what is sold as 'reality' now.

But back to my story (sorry for the erratic style of this message; cerebral upgrades have become too expensive thus my version is outdated; my memory is falling apart).

In a society that is turned to maximum material growth, where success is tied to delivery, not getting an upgrade quickly equals being an outcast. Performance pays. Markets decide how much and what for. With companies fighting to survive and expand, most small and medium size businesses were suffocated or swallowed by sequels of the former FAAMG corps. Workers take what they can get; as production processes are completely automated jobs are rare. Human replacements occurred successively; first manual labor like factory floor work was eliminated; then caretaker jobs, teachers, nurses, etc.; finally doctors and artists became obsolete. Doctors followed because AI was less error-prone and more precise, artists because there was no demand. Since no one can now distinguish between human and machine generated books, paintings, movies, music, etc., no one is willing to pay for genuine artifacts.

The only ones who retained employment are the engineers who serve as a backup, in case the integrated repair mechanisms of our robotic workforce fell out. As well, those who pull the strings of power stay. I am not talking about politicians—that is another outdated job-profile that eventually left the scene; but CEOs. (When I was a child the pretension still prevailed, that society should be ruled by ideas and ideals, merit; that those who oversaw countries should be elected democratically. But eventually the concepts of power and money merged. Democracy, kindness, altruism, and various other concepts did not make it to your time. I do suggest you check out the aforementioned website to get an idea of alternative forms of social cohabitation.) In the past money was, at least theoretically, one element among others that determined a person's social status. Now money, mind and meaning are synonymous.

Even though brain-chips have gone mainstream, access remains limited. As before education, food, health care and decent employment, brain-grading joined the list of 'commodities' that determine the quality of our life. These commodities are not accessible to everyone. The imbalance between needs and means is bigger than ever before in humankind. Because whether and what type of upgrade you can afford is a question of money. In return the sophistication of your upgrade influences your success and status in society. It is a vicious cycle which continues to accelerate. The cycle of upgrades becomes ever shorter, while the range of offers expands. Billions around the world are thus taken hostage by debts, incurred in the hopeless attempt of remaining up to date. Sure, everyone gets the basic obligatory routine upload, which has replaced primary education and secondary education. But already the content of this basic set-up is marked by inequality. Quality standards differ widely, not only between countries (some are off the charts) but even within countries the high grades are reserved for those who are part of the 1% purchasing power peak.

Overall it seems fortunate that birth rates have shrunk drastically over the past decades. Robots can fornicate but are not fertile. And though people do not have to work to make a living, nobody has time or volition to invest in relationships. Why make the effort involved with finding, grooming, and entertaining a partner when a robot looks, feels, talks, and behaves like a human would; without cumbersome opinions, moods, and needs. Once the norm, offline human2human connections have become an exception. MachineMating is the new normal now. I am telling you this because I want you to know that there is an alternative—people used to meet and mate, talk and touch. The current trend is reversible, if individuals like you choose to act.

I did not do enough—and maybe this letter is the most useful thing I have ever done. Please let me go trusting that you will choose your own life; that you will not accept an existence attributed by AI.

Take care of yourself, and the planet that you are part of. Always remember, everything is connected.

I understood that too late.

Yours forever. A

PS: I mentioned the POZE paradigm. It still exists. Thanks to small islands of like-minded thinkers and doers who resisted the mainstream.

You can reach them via https://www.poze.cc Give them, and yourself, a fair chance. Trust your path, not machines.

****

Futurist Letter 2—Aspiration Ahead (+)

My Love,

I am writing these words because I want you to treasure your being and becoming, every day.

You are born in extraordinary circumstances. What appears today as normal was once not only unimaginable, but undesired. People sought refuge in offline and online escapism, in the hopeless attempt to numb the call for meaning. From my parents I heard gruesome stories about violence and poverty, discrimination and inequality. Concepts that seem outdated these days, because their detrimental impact is not only obvious (which was apparently also the case then), but because the underlying causes that triggered them then are nowadays shunned by Society.

In hindsight it is challenging to pinpoint when the shift from darkness to light began. But historians mostly agree that the Pandemic which engulfed the World in 2020/2021 triggered a first wave of drastic change. It pulled the veil of a conundrum of social inequality that had been perpetrated for centuries, where many had too little, and a few had too much. It sounds counterintuitive but this imbalance was taken as a given. People considered it as normal that not everybody could, or should, be safe from harm. They accepted that billions of individuals, people like you, me, and themselves, were deprived of healthy food, quality education, clean water, safe shelter, and enjoyment. Those whose needs and wants were covered perceived no push to change the prevailing imbalance of needs and means. They did not see the link between their own interests and that of others, and their logic was entirely condoned by society.

You have grown up with the understanding that giving is having and sharing means growth. Seeing that these are key features of your education you are by now well acquainted not only with the concept of quality of life (QoL), you are aware that it is not the exclusive privilege of some people, but the entitlement of everyone. This normalization of compassion and generosity is probably the biggest difference between you and your forefathers and mothers. As your LCC* has explained to you from the first session, respect and responsibility for us, and others underpins who we are and become.

Excuse me for boring you with aspects that are obvious. I just want you to be acutely aware that these shifts in human consciousness are recent, and that they represent a landmark in human history that must be protected. Never take them for granted!

Remember that the AA** that helps us achieve the best possible outcome for people and planet, while minimizing the cost of basic QoL components such as food, water, shelter, health care, education, transportation, communication, etc., were designed by human engineers. Technology is all pervasive today; but it was envisioned, designed, and delivered by individuals like you and me. It is to maintain that level of creativity that you have been taught early on to spend time off the grid, to contemplate, dream, play, and aspire to be transformational. As you know, the impact of our technological helpers derives on the one hand from the intentions that underpinned their creation, and on the other hand from the use that we make of them. Cause and consequence are clear.

Never consider the system that you evolve in as an excuse for something that happens to or around you. The past illustrates that systems can always change. Ideally such change occurs organically, before the existing system is outdated. Alternatively, sudden shocks cause the onset of radical upheaval. Following centuries of deprivation, the death of billions of people from futile causes like hunger, violence, disease, etc., 2021 brought such a sudden shift, due to the combination of fallouts from COVID-19, chronic deprivation, and a global crises of purpose void. It caused a critical mass of people around the world to wake up and step forward to act. That shift started a trend toward systemic understanding that is felt until today. Over the years most individuals and institutions adopted the holistic perspective that you were taught in kindergarten. Their 360-degree perspective underpins the global optimization logic that you have grown up with.

By the 2040s all countries had adopted some sort of universal basic income scheme, because they realized that compensating unemployment with selective social services is inefficient. International funding transfers made this possible globally, within a coherent organically revolving master-matrix that covers all nations; reaching every human being.

As the link between money and labor has dissolved you may never 'have to' work to make a living. In the past many people wasted their life on tasks that were not only disconnected from their passion but detrimental to their health, their family, and the environment. Our forefathers spent the lion-part of their lives to make money to buy stuff. When they reached their 60s, they retired. By then their physical mental condition had begun to decline. Most of the stuff they had spent their energy on was gone or junk, and so they spent whatever time and money they had left to survive and keep busy. Many of them lived and died alone because they had no time to cultivate friendships. This dark chapter of humanity is hopefully behind us.

If the current evolution of humanity continues to progress—not forward but up, in a holistic understanding of spiritual and emotional progress, then the gloomy concepts that I referred to above will drift ever further away as time goes by. I am optimistic. The introduction of generosity, compassion, honesty and courage in school curricula around the world has led to the normalization of purpose-oriented behavior. Your generation is growing up to be the foundation of a radically different collective mindset. Your children may reach the next ZENITH, individually and together.

Always remember—Everyone is unique. Nobody has the same combination of skills and resources as you do. Thus, nobody can fulfill the mission that you have come to accomplish in this life. If you do not identify and pursue your purpose a void remains in this World. You learned it from your parents and your LLC is reminding you every day: Your WHY is your compass. You must never expect technology to replace it.

Love, B

*Life Cycle Coach
**Aspirational Algorithm

NOTES

1. A variation on the term, "garbage in, gospel out," refers to a tendency to put unwarranted faith in the accuracy of computer-generated data (Rouse 2008).
2. The Green Revolution (also called the Third Agricultural Revolution) refers to a set of research technology transfer initiatives occurring between 1950 and the late 1960s, that increased agricultural production worldwide. The initiatives resulted in the adoption of new technologies, including High-Yielding Varieties (HYVs) of cereals. It was associated with chemical fertilizers, agrochemicals, controlled water-supply and newer methods of cultivation, including mechanization. Together these 'technologies' were offered as a package of practices to be adopted

together as a way to supersede 'traditional' technology (Farmer 1986). The impact on the environment was detrimental in many cases. Today more than enough food is produced to feed the global population—but more than 690 million people still go hungry (FAO 2020). Indeed, after steadily declining for a decade, world hunger is on the rise since 2014; affecting 8.9% of people globally. Conflict is a major driver of hunger. Conversely the latter is a major cause of migration and civil unrest. Everything is connected, and triggered by human action.

3. For an overview of the most common biases see Desjardin (2018).

4. Contact tracing identifies people who were recently in contact with an infected individual, in order to isolate them and reduce further transmission. Digital tools for contact tracing may be grouped into three approaches: (1) outbreak response; (2) proximity tracing; and (3) symptom tracking. It serves to augment and accelerate manual contact tracing (Anglemyer et al. 2020).

5. Being grateful has benefits in each of our 4 dimensions, and smoothens the interplay between them, as well as ourselves and others. Benefits can be grouped in five groups (Ackermann 2020): Health, Emotional, Personality, Social and Career benefits.

6. Reinforcement strengthens an organism's future behavior whenever that behavior is preceded by a specific stimulus. Whereas positive reinforcers provide a desirable stimulus; a negative stimulus includes a sort of (perceived) punishment (Schultz 2015). Similar to Priming, Reinforcement does not require an individual to consciously perceive an effect elicited by the stimulus. (Winkielman et al. 2005).

7. The Hawthorne effect concerns research participation, the consequent awareness of being studied, and possible impact on behavior. A large literature and repeated controversies have evolved over many decades as to the nature of the Hawthorne effect. Despite the heterogeneity of results the overall picture that emerges indicates that a causal link between (the perception of) observation and behavior exists (McCambridge et al. 2014).

8. An early illustration of systemic thinking for social solutions was Forrester's (1971) model of world dynamics which correlates population, food production, industrial development, pollution, availability of natural resources, and quality of life, as well as projections of those values into the future under various assumptions. It served as the initial basis for the World3 model used by Donella and Dennis Meadows in 'The Limits of Growth' (1972). The latter was criticized extensively by many economists; some describing it as "an empty and misleading work [...]" (Passell et al. 1972). Nonetheless, four decades later a study conducted by researchers at the University of Melbourne found the forecast to be accurate (Turner 2014).

9. Bernoulli's Hypothesis states a person accepts risk not only on the basis of possible losses or gains, but also based upon the utility gained from the risky action itself. His hypothesis introduced the concept of expected utility, stating that the amount of utility from playing a game is a significant decision factor in whether or not to participate.

10. Bounded rationality is the idea that rationality is limited, due to the tractability of the issue, the cognitive limitations of the human mind, and the time available to make the decision. "Decision-makers, in this view, act as satisficers, seeking a satisfactory solution rather than an optimal one" (Simon 1955). People are not able/willing to make a full cost-benefit analysis that would determine the optimal decision; they choose an option that fulfils their adequacy criterion. Humans take reasoning short-cuts that may lead to sub-optimal decisions. As seen in Chapters 1 and 3 nudges and intentionally shaped decision-making architectures as proposed by behavioral economists may serve to enhance the outcomes of these choices. However, as noted earlier the pursuit of an optimum is the quest of a vision. It is motivating yet whichever 'optimized' state is reached, it will be termporary and a transition.

11. In a non-linear, ever changing world, there is no optimum, only satisfaction and constant (or fluctuating/cyclical) improvement along a path-dependent search (Simon 1976).

12. On the ensuing moral dilemma see Nelson (1977).

13. "We have developed speed, but we have shut ourselves in. Machinery that gives abundance has left us in want. Our knowledge has made us cynical. Our cleverness, hard and unkind. We think too much and feel too little. More than machinery we need humanity. More than cleverness we need kindness and gentleness. Without these qualities, life will be violent, and all will be lost...." From Chaplin's closing speech in the 1940 movie "The Great Dictator", see Chaplin (1964).

REFERENCES

Ackermann, C. (2020). 28 benefits of being grateful & Most significant research findings. *Positive Psychology*. Retrieved March 2021. https://positivepsychology.com/benefits-gratitude-research-questions/.

Anglemyer, A., Moore, T. H. M., Parker, L., Chambers, T., Grady, A., Chiu, K., et al. (2020). Digital contact tracing technologies in epidemics: A rapid review. *Cochrane Database of Systematic Reviews, 2020*(8). Art. No.: CD013699. https://doi.org/10.1002/14651858.cd013699.

Barrett, L. (2019). *You aren't at the mercy of your emotions—your brain creates them*. TED. Retrieved November 2020. https://go.ted.com/CdJA.

Bicchieri, C. (2016). *Norms in the wild. How to diagnose, measure, and change social norms*. Oxford: Oxford University Press.

Bohm, D. (1980). *Wholeness and the implicate order*. Routledge. ISBN 0-203-99515-5.

Brynjolfsson, E. (2013). *The key to growth? Race with the machines*. TED. Retrieved October 2020. https://go.ted.com/6b7a.

Cannon, W. (1932). *Wisdom of the body*. New York: W.W. Norton. ISBN 978-0393002058.

Carr, N. (2010). The internet scatters focus rewires brains. *Wired*. Retrieved August 2020 from https://www.wired.com/2010/05/ff-nicholas-carr/.

Chaplin, C. (1964). *My autobiography Charles Chaplin 1964 hardback by Charles Chaplin (1964-08-01)*. Simon & Schuster.

Desjardin, J. (2018). *Here are 24 cognitive biases that are warping your perception of reality*. https://rb.gy/urwvvy.

Dewey, J (1938) *Experience and education*. Free Press; Reprint Edition (July 1, 1997).

Farmer, B. H. (1986). Perspectives on the 'green revolution' in South Asia. *Modern Asian Studies, 20*(1), 175–199. https://doi.org/10.1017/s0026749x00013627.

Food Agricultural Organization (FAO). (2020). *State of food insecurity and nutrition in the world 2020* online summary. Retrieved November from http://www.fao.org/state-of-food-security-nutrition/en/.

Forrester, J. (1971). Counterintuitive behavior of social systems. *Technology Review, 73*(3), 52–68.

Ghaffary, S. (2019). *Many in Silicon Valley support universal basic income. Now the California Democratic Party does, too*. Retrieved October 2020. https://rb.gy/9h09sr.

Gilbert, D. (2005). *Why we make bad decisions*. TED. Retrieved July 2020. https://rb.gy/g9qdez.

Gladwell, M. (2007). *Blink: The power of thinking without thinking*. Back Bay Books.

Hawkins, S. (1988, September 1). *A brief history of time*. Bantam; 10th Anniversary edition.

Heiner, R. A. (1983). The origin of predictable behavior. *The American Economic Review, 73*(4), 560–595.

Hills, P., & Argyle, M. (2001). Happiness, introversion–extraversion and happy introverts. *Personality and Individual Differences, 30*(4), 595–608. ISSN 0191-8869. https://doi.org/10.1016/S0191-8869(00)00058-1.

Kahneman, D. (2011). *Thinking, fast and slow*. London: Penguin Books.

Kompanje, E. J. O. (2018, May). Burnout, boreout and compassion fatigue on the ICU: it is not about work stress, but about lack of existential significance and professional performance. *Intensive Care Medicine, 44*(5), 690–691. https://doi.org/10.1007/s00134-018-5083-2. Epub 2018 February 21. PMID: 29464299.

Matthews, D. (2006). Epistemic humility: A view from the philosophy of science. In J. P. Van Gigch & J. J. McIntyre-Mills (Eds.), *Wisdom, knowledge, and management: A critique and analysis of Churchman's systems approach* (p. 113). New York: Springer (1959).

McCambridge, J., Witton, J., & Elbourne, D. R. (2014). Systematic review of the Hawthorne effect: new concepts are needed to study research participation effects. *Journal of Clinical Epidemiology, 67*(3), 267–277.

Miller, R. (2020). *Garry Kasparov on AI: 'People always called me an optimist'*. TechCrunch.

Nelson, R. (1977). *The Moon and the Ghetto: An essay on public policy analysis.* New York: W. W. Norton.

Nelson, R. R., & Winter, S. G. (1982). *An evolutionary theory of economic change.* Cambridge, MA: Harvard University Press.

Otake, K., Shimai, S., Tanaka-Matsumi, J., Otsui, K., & Fredrickson, B. L. (2006). Happy people become happier through kindness: A counting kindnesses intervention. *Journal of Happiness Studies, 7*(3), 361–375.

Passell, P., Roberts, M., & Leonard, R. (1972). The limits to growth. *New York Times.* Retrieved December 2017.

Post, S. (2014). It's good to be good: 2014 biennial scientific report on health, happiness, longevity, and helping others. *International Journal of Person Centered Medicine, 2014*(2), 1–53.

Pressman, S. D., Kraft T. L., & Cross, M. P. (2015). It's good to do good and receive good: The impact of a 'pay it forward' style kindness intervention on giver and receiver well-being. *The Journal of Positive Psychology, 10*(4), 293–302.

Ricard, M. (2004). *The habits of happiness.* TED. Retrieved July 2020 https://rb.gy/iqrjf9.

Rouse, M. (2008). *Definition garbage in, garbage out.* SearchSoftware. Retrieved November 2020. https://rb.gy/tygl0h.

Schultz, W. (2015). Neuronal reward and decision signals: From theories to data. *Physiological Reviews, 95*(3), 853–951. https://doi.org/10.1152/physrev.00023.2014.pmc4491543.

Schwartz, B. (2004). *The paradox of choice—Why more is less.* New York: Harper Perennial.

Simon, H. A. (1955). A behavioral model of rational choice. *The Quarterly Journal of Economics, 69*(1), 99–118. https://doi.org/10.2307/1884852.

Simon, H. A. (1976). From substantive to procedural rationality. In T. J. Kastelein, S. K. Kuipers, W. A. Nijenhuis, & G. R. Wagenaar (Eds.), *25 years of economic theory*. Boston, MA: Springer.

Stairs, M., & Galpin, M. (2010). Positive engagement: From employee engagement to workplace happiness. In P. A. Linley, S. Harrington, & N. Garcea (Eds.), *Oxford library of psychology. Oxford handbook of positive psychology and work* (pp. 155–172). Oxford University Press.

Swah, M. (2019). *Tapping into the brain: Neuroscience wearables explained.* Retrieved July 2020. https://rb.gy/vzqpty.

Thaler, R., & Sunstein, C. (2008). *Nudge: Improving decisions about health, wealth, and happiness*. London: Yale University Press.

Turkle, S. (2012). *Connected but alone*. TED. Retrieved November 2021. https://www.ted.com/talks/sherry_turkle_connected_but_alone.

Turner, G. (2014). *Is global collapse imminent?* Melbourne University. Retrieved November 2020. https://sustainable.unimelb.edu.au/publications/research-papers/is-global-collapse-imminent.

United Nations. *Human Rights*. Retrieved August 2020. https://www.un.org/en/sections/issues-depth/human-rights/.

Von Hippel, W. (2018). *The social leap: The new evolutionary science of who we are, Where we come from, and what makes us happy*. New York, NY: HarperCollins.

Walther, C. (2020b). *Humanitarian work, social change and human behavior*. New York: Palgrave Macmillan.

Walther, C. (2020c). *Connection in times of Covid. Corona's call for change*. New York: Palgrave Macmillan.

Wickstrom, G., & Bendix, T. (2000). The "Hawthorne effect"-what did the original Hawthorne studies actually show? *Scandinavian Journal of Work, Environment & Health, 26*(4), 363–367.

Winkielman, P., Berridge, K. C., & Wilbarger, J. L. (2005, January). Unconscious affective reactions to masked happy versus angry faces influence consumption behavior and judgments of value. *Personality & Social Psychology Bulletin, 31*(1), 121–135.

Wright, E. O. (2016). *Can the universal basic income solve global inequalities?* World Social Sciences Report 2016. UNESCO. Retrieved March 2021. https://shorturl.at/citS8.

INDEX

C. C. Walther, *Technology, Social Change and Human Behavior*, https://doi.org/10.1007/978-3-030-70002-7

The manufacturer's authorised representative in the EU is Springer
Nature Customer Service Centre GmbH, Europaplatz 3, 69115 Heidelberg,
Germany. If you have any concerns regarding our products, please
contact ProductSafety@springernature.com

Printed and bound by CPI Group (UK) Ltd, Croydon, CR0 4YY
24/04/2026
02096337-0001